W

Twenty Five Years Of
The Austin Maxi

Selwyn Daley

Berserk Publishing
1994

Published by **Berserk Publishing**, Cultural Centre, Redhill Avenue, Glasshoughton, West Yorkshire, WF10 4QH

© Selwyn Daley and Contributors
© Poem on page 32: John Hegley

© Cover Photograph: Ian Brook of Weymouth
© Cover Design: Rich Whale of Ergo Design

Typeset and printed by Berserk Publishing

ISBN: 1 899661 00 X

Classification: Motoring

Thanks to:
Isabel Galan, John Salt, Richard Rouska, Reini Schühle, Olive Fowler, Steve Davenport, Brian Lewis, Austin Maxi Owners Club, Yorkshire Art Circus

Special thanks must go to my girlfriend Isabel, who has bailed me out of trouble on several occasions during the course of putting this book together.

For details of the Austin Maxi Owners Club contact:
Maxi Owners Club
27 Queen Street
Bardney
Lincolnshire
LN3 5XF

British Library in Publication Data. A catalogue for this book is available from the British Library.

All rights reserved. No part of this publication may be reproduced or transmitted in any form or by any means, electronic or mechanical, including photocopying, recording or any other information storage and retrieval system without permission in writing from the publisher.

Contents

Introduction   6

Factor One: Character   10
The Maxi, The Frying Pan
and George The Pig

Factor Two: Practical   32
The Car For People With
More Sense Than Money

Factor Three: Holiday   48
I Wanna Tell You A Story
Tour-A-Long-A-Max

Factor Four: Performance   70
Brake On, Brake Off
Break Up, Breakdown

Factor Five: Bonding   94
Part Of The Family
In Less Than A Week

Epilogue   112

Contributors

H R Allen
Jeff Allen
N D Amis
John Arnold
H M Ashman
Hilda Ashton
Tom Ashworth
Chris Atkin
Margaret Baker
Tim Bannerman
Barbara Beale
Louis J Beddall
John Beddell
David Boardman
Howard Brooksbank
Canon Derek Brown
Geoff Brown
John Brown
Harry Butcher
Brian Butterworth
F P Cain
E J Carpenter
Tim Carroll
Alison Carter
Robert Corden Champ
H Checkley
Louise Chilton
D Claridge
Philippa Clark
Ade Cook
Morry Cook
Liz Croft
George Currie
Steve Davis
Ian Dees
Ken Elshaw
Nigel Flynn
Dick Foot
Betty Forsythe
Katy Fox
John Fraser
E French
Isabel Galan

R Gatehouse
Sean Geraghty
Roy Gillespie
P T Grantham
M E Gregory
I G Hale
Keith Halsey
A Harper
Paul Harris
Roger Hart
Graham Haworth
J Hay
John Hegley
David Henshaw
Geo Hill
Nairn Hindhaugh
Clayton Hirst
Ian Holdsworth
John Howe
Stanley Hutton
A Ingham
Christine Jackson
Phil James
Chas Jenkins
Noel King
S Korbey
Peter Lawley
J D Legge
Neil D Lomas
Don Louth
Phil Malby
Philip Marrison
Helen Mason
Kirsty McNicol
R McNicol
G T Mills
P Morgan
G Mulveny
George Mulveny
Jan Nelder
C I Old
C F Old
J L Osmond

Des Penny
Anne Pilcher
Robin Pilcher
Molly Pollitt
James Prickett
Joan Renton
Pamela Reveirs
Dave Richards
Paul Richards
Bob Ritchie
Donald Robinson
David Rowe
C Samuel
Hugh Sayce
Keith Sharpe
Neil Shirley
Ian Simmons
Brenda Collis Smith
David Collis Smith
John Sommerschield
Peter Sothcott
Geoff Speight
John Startin
Roger Statham
Catherine Stratford
Robert Stratford
Stephen Tames
Andrew Tarling
Tony Teague
Ian Tennant
David Tinsley
D Treble
George Tripp
Rev G H Turner
Ron Watts
P A Wayman
Kenneth Wells
J Wheeler
Edwin White
Mr Whiting
Geoff Woods
Rachel Yapp

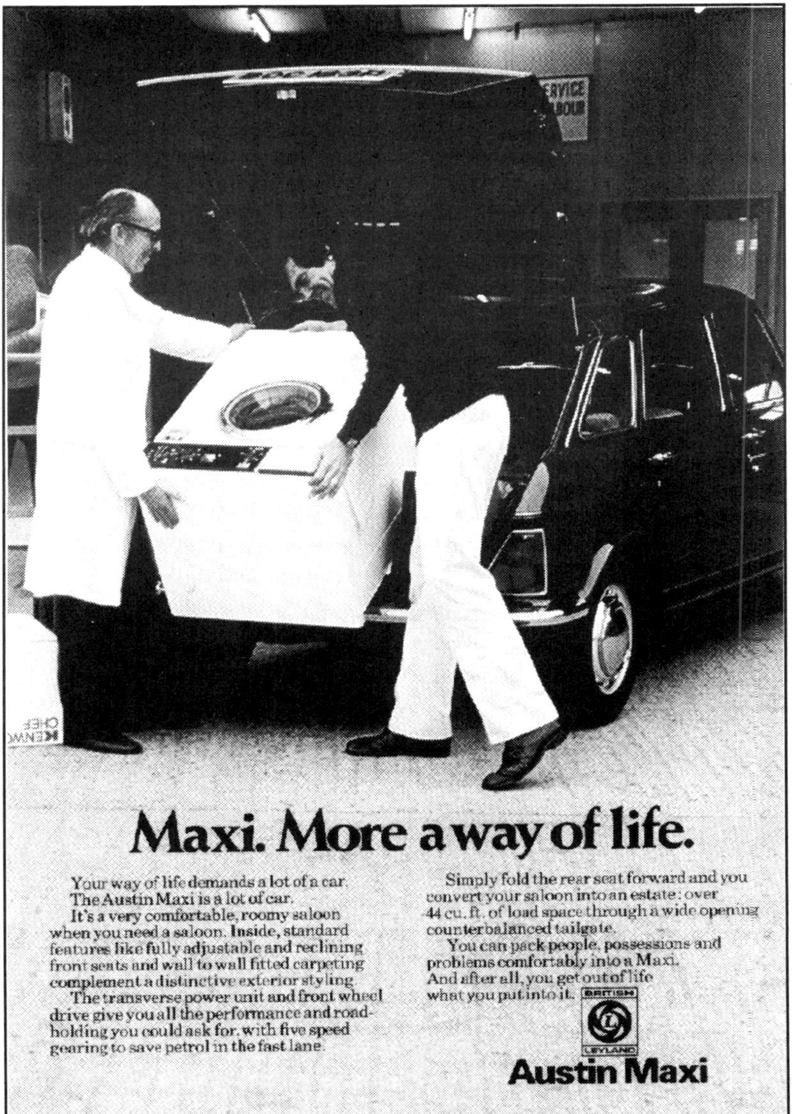

Maxi. More a way of life.

Your way of life demands a lot of a car. The Austin Maxi is a lot of car.

It's a very comfortable, roomy saloon when you need a saloon. Inside, standard features like fully adjustable and reclining front seats and wall to wall fitted carpeting complement a distinctive exterior styling.

The transverse power unit and front wheel drive give you all the performance and road-holding you could ask for, with five speed gearing to save petrol in the fast lane.

Simply fold the rear seat forward and you convert your saloon into an estate: over 44 cu. ft. of load space through a wide opening counterbalanced tailgate.

You can pack people, possessions and problems comfortably into a Maxi. And after all, you get out of life what you put into it.

Austin Maxi

Introduction

I love the Austin Maxi. It holds a mysterious charm that is essentially ridiculous, but has a beauty that it carries like no other machine could. It is a righteous, simple, kind-hearted, uncorrupted, scatty, bona fide freak of engineering that I embrace with a passion. For the last four years I have been captivated by its charm; I've joked about it, I've bragged about it, I've written songs about it. I have somehow managed to infiltrate the Maxi concept into everything I do.

I write a regular column for the *Wild Rover*, the Rugby League fanzine of my hometown club, Featherstone Rovers. Each issue, I indulge in the Maxi, the fortunes of the club are mirrored by the fortunes of my car and I frequently put out a call to the fans to chant 'Austin Maxi' at the opposition in an attempt to unnerve them. In my role as a performer in the comedy double act, *Selwyn and Dick*, I throw in several references to the Maxi and in one bizarre routine we put Gandhi in the back of one.

My infatuation with the car has often puzzled me. To me, the Maxi is just one big bundle of fun and it gives me status, but I cannot escape the fact that it can easily be viewed as nothing more than a big cube of pressed steel, pulled by an inconsistent engine and belongs stylistically to a not too distant, uncelebrated age. Until very recently, I thought I was the only person who paid the Maxi much attention.

In December, 1993, I got a phone call from an old college friend, John Salt. He'd called to tell me that he'd got a job with the *Guardian* newspaper selling advertising for the motoring pages. He said he would like to impose his personality on the page and liven up what is a very static lump of space. He wanted me to advertise my Maxi for sale at a very low price in a national broadsheet newspaper. I laughed, it was a funny notion, but I was quick to realise that I was being offered very cheap national advertising and that kind of chance didn't present itself everyday. I drew on a cigarette, and the result of whatever chemical reaction took place when the nicotine hit my brain, you are holding in your hand.

'Why don't we go for a book?'
'What?'
'Wanted, anecdotes for inclusion in a book on the Austin Maxi.'

And so it was, on Christmas Eve an advert appeared in the *Weekend Guardian*. That was far as I expected it go. I'd thrown out the bait to see if there were any takers but, realistically, I talked myself down to expecting no more than the odd reply out of curiosity. I had no concept that anybody else out there carried a candle for the Maxi in quite the same way as I did.

When I walked into my office following the New Year's holiday, I couldn't believe it. There were ten letters on the doormat and a string of messages on the answer machine. I had struck gold. Many of the letters expressed sentiments of isolation and that they thought they were the only ones too. At last a chance to get something together.

I telephoned another college friend who was working as a personnel officer for the Rover Group in Cowley. I was hoping she might be able to put her hands on some Maxi leads. She rang back within the hour.

'There is an Austin Maxi Owners Club.'

'You are kidding.'

I rang up Christine Jackson, the secretary of the club. It was true, they had 350 members, a regular magazine, they meet at least twice a year and by the way, did I know it was the Maxi's twenty-fifth birthday. Of course I didn't.

This was the answer. If I could hear from all these people and get their story, maybe I might get close to pinning down exactly what the attraction is. I put a page in the next copy of *Maximise*, the club's magazine, and John Salt, who has shared my enthusiasm for this project all the way through, managed to squeeze the advertisement into the *Guardian* and *Observer* whenever he could.

So the stories kept coming. For the whole of 1994 a week hasn't gone by without receiving a letter from someone telling me all about their adventures in a Maxi. The diversity of the people has been astonishing. There has been mail from a headmaster, a dentist, an ambulance driver, students, professionals, a few men of the cloth and even one of the all time greats of snooker. It was nice to know I wasn't alone in my love of the Maxi, but the cross-section of the public who drive them and get pleasure for all sorts of different reasons only served to confuse the issue of attraction. I couldn't imagine a dentist wanting a cheap run about; it didn't fit that a headmaster would like the thrill of blasting down the motorway on a knife edge, never knowing if he was going to get past the next car or spontaneously combust and I just presumed that men of the cloth were not seduced by the fact that you can fold the seats down to make an impromptu double bed, should the need suddenly arise.

We do spend a lot of time penned up in our cars. I have heard it said many times that a car is an extension of the personality and I must say, I tend to agree with that. In moments of insecurity, when I translate that notion into my own situation, I get terribly anxious. Not only do I drive this contraption, but I babble on about it at great length and considering the public conception of the car is not very favourable, if I'm not careful I can quickly spiral down into depression. On good days I'm confident that the Maxi is noble, and that owning and loving one gives me a feeling of exclusivity, which is a special buzz to anyone.

I am aware of the two forces surrounding the Maxi. There is the general attitude of condemnation and the much smaller informed opinion that it is extraordinarily special. I began the long task of sifting through my mail and trying to find some sort of order in hope that, like a jigsaw puzzle, when all the pieces are put in the right place, there to behold is a beautiful and very clear image of what you have been working towards.

I have been in contact with well over one hundred Maxi people during the last twelve months. They have given me information that ranged from a two minute conversation on the telephone to a ten page epic on the exploits of their own particular Maxi. A lot of the material I received has found it's way into the book, but there was a lot of repetition so not every story is here. However each piece of information I have received has fed into the final editing decisions and I have in no way tried to distort the picture I was being sent.

I have done a substantial amount of research into the Maxi through other publications, but found very little of any substance. There is a photocopied brochure by Trevor Adler of Transport Resource Books that pulls together all the Maxi related stories from the motoring press. Although a fascinating read, it was mainly technical so I didn't fully understand what most of the articles were on about. There perhaps should be a book written about the mechanical niceties and quirks of the Maxi, but this isn't it and I am certainly not the one to write it.

I know virtually nothing about the mechanics of the car, I leave all that to an old school friend who runs *Graham and Press Autos* in Featherstone. What he doesn't know about Maxis by now, isn't worth knowing. One morning I was surprised when Phil Malby, whom I'd never met before, rolled into the car-park of where I work and casually said that he thought I might like to see his Maxi. It was a very nice one, registered on 1st May 1969, the very first day Maxis were sold. I was impressed by its body and its condition but when he insisted on opening the bonnet to proudly show me the original starter motor and alternator, I had to ask him to point. It is that bad.

This book is about the people who drive Maxis, it is not about the car itself. It would not have helped me in my quest to find out what the attraction of the car is by finding out the best setting for the points or the length of the exhaust.

'I like the Austin Maxi because you set the points to forty-eight thou and that is very unusual in a car.'

It isn't the answer. I wanted to know what happened to the car. I've read the company line on what they wanted it to be, I've read the advertisement copy that listed the selling points of the car and I've read the motoring press version that says what it probably can't do. What I wanted to hear was what it has actually meant to the public over the last twenty-five years and how it has affected their lives.

The stories I have received have been amazing. When I put them all together, a pattern did emerge. I decided that there are five strands to the attraction of the Austin Maxi; character, practicality, the holiday factor, performance and the bonding factor. I shuffled all the stories into their various sections, you can see in the result in the following pages.

Each section is in two parts. The first is an introduction from me, off-loading all the relevant things that have happened in my Maxis. The second part is the contributors' section, containing the stories that have been sent in during the course of the year.

One gentleman called me and warned me against producing a 'hippy book'. He said that the Maxi was a fine car and received enough bad publicity over the years and that I should leave it alone. Now I have no idea what a 'hippy book' is, so I can't say if this is one or not. What I can say is that the purpose of this book is only to celebrate. The Austin Maxi is unquestionably unique and I see this book as a fitting tribute to it.

Character

My trainers say I'm trendy
My tank-top says I'm old
My straw hat says I'm boiling
My long-johns say I'm cold
But there's nothing can be gotten
I become a mystery
When I jump behind the wheel
Of my Austin Maxi

It is not in dispute that the Maxi is full of character. I have been in many situations where the presence of my Maxi has added an extra absurd dimension to what are already outrageous events. Recounting many stories, I often put the Maxi as the central character - it is guaranteed to bring bigger laughs. I was pleased to see in the mail I received that I wasn't the only one.

A few years ago I was contracted as a slave for a prestigious London based catering company that held a number of Royal Crests. We fried eggs in some first class venues. It was a case of the 'Tuesday, must be face down in a dustbin at Kensington Palace; Wednesday, must be a bun fight at Guildhall' type of three pound per hour freewheeling lifestyle.

I managed to escape that career path with two charming memories. The first event occurred at Highgrove House when we were setting up an outdoor event for the Prince's Trust. Out of nowhere appeared two little boys on mountain bikes, the eldest of whom asked, 'Can we be of any help?' It was the two young princes, William and Harry.

In for a penny, I stood back, lit up a cigarette, assumed a supervisory position and said, 'You see those tables over there, they need to be over here.'

Eager to please, they went to work. A touch of 'left a bit, right a bit' from me and the tables were moved in no time.

'Don't forget the chairs, if a job's worth doing...' I was enjoying myself. They did as they were asked and were off.

A week later I was carrying some bone cracking, back breaking beer crates through a marquee and as I paused to wipe the sweat pouring from my brow, I glanced at a television set and saw the young princes being escorted around the starting grid at the British Grand Prix with Jackie Stewart. It was a powerful moment.

The second magical memory from that time with the cooking crew was far more prestigious and a touch more relevant. I got my Maxi into the grounds of Buckingham Palace. Well, almost.

It was a scam from start to finish. We were setting the equipment up in preparation for the summer garden parties and had to start quite early in the morning. I was living in South-East London at the time and the journey in wasn't a pleasant one. I've never been fond of mornings, and jumping out of bed onto a train and then another train just put me in a bad mood. I suggested to my manager that I would like to bring my car in and park it in the Palace grounds. He laughed and said he couldn't even get his in, wasn't I aware of the class system and that if there was a space he would be some distance in front of me in the line. I was deeply unmoved.

As I was leaving the Palace that day, I had a quiet word with the policeman on the gate and explained the early morning, two train, bad mood blues and was there any chance of getting my car in. He seemed to think there wouldn't be a problem, but I should contact the PC on duty in the morning. He gave me a telephone number and told me to ring in before I set off.

I tried, but I couldn't conceal my delight. I headed straight for the nearest pub, the Bag O' Nails on Buckingham Palace Road, drank a toast to Austin Maxis and called for the head of my boss.

The following morning I rang the Palace as instructed and spoke to a very positive police officer.

'What kind of car is it sir?'

'Austin Maxi.'

'No problem, I will be waiting for you.'

It was the most pleasant drive through London I have ever experienced, green lights, gaps in junctions, clear high streets and the Maxi was cruising. I arrived at the Palace in no time, explained who I was and watched as the big gates slowly swung open. I was in.

At this point I should describe the state of my Maxi. It was coming to the end of its life, there was no exhaust to speak of, the bodywork was falling to pieces, the aerial was a coathanger shaped like a fish and I'd painted 'Selwyn' in big black letters on the driver's door. If it was alright for Tom Cruise in *Days Of Thunder* then it was alright for me.

Once inside the main gates, I was confronted by a second set, big iron ones, staffed not by my friendly policeman, but a military-looking Palace official. He could sense a Trojan horse and was determined not to be the idiot who made that mistake again. Entry was refused and I was asked to leave.

A bigger stink I could not have caused. I had followed procedure, I was working in fifteen minutes and I was in the centre of London with no alternative parking. My Maxi wasn't going anywhere.

A gathering of heavy officials and heavy words were used to remind me of the weakness of my position. They were an army, I was one man with a clapped-out Maxi. But I sensed injustice, I had done the right thing and I'd been misled. The romance of getting my Maxi into the Palace had taken me over and I was prepared to go to the Tower for it. I demanded the policeman to stand up for me.

A compromise was reached. Across the road was an exclusive underground parking lot, with three unaccounted spaces. If I agreed to keep quiet I could park there today and the rest of the week. In protest I agreed, turned the ignition - nothing. I'm sure it would have sarted if it wanted to but it didn't. The soldiers had to bump me off.

I crossed the road, into the underground lot and parked up between a Ferrari and a classic Bentley. I was carrying a footpump at the time due to a slow puncture in one of the front tyres. Each day after my shift at the Palace, I would have to get it out and do some severe pumping. I often wondered what the other owners thought when they witnessed this act in the knowledge that a day's rent on that lot probably cost more than my entire worldly possessions.

There is a footnote to this story. Six months later, we were booked again to work at the Palace to do the Household's staff Christmas party. As I went to collect my security pass, the uniformed gent who dealt with me said, 'I remember you, Austin Maxi.' I was let in with a nod for the length of the job, no pass, no search, nothing.

It struck me that if I had driven any other car I wouldn't have received the same treatment. I couldn't imagine the official saying, 'I remember you, Renault 18,' it just isn't right. The Maxi had given me a certain credibility. In this instance I think the old British imperialist side of the Maxi came to the fore. There I was with a motor car that tipped a nod in the direction of the modern world and was aware of foreign competition, but was essentially a symbol of all things British. I was at the gates of our monarch's house trying to get in. I was looked upon as one of them, I had unwittingly drawn a parallel with my car and the state of the nation. The things you can do with a Maxi.

In other circumstances a different side of the Maxi comes to my assistance. When I pull up in the car park of an exclusive wine bar, coughing, banging and spluttering next to pristine BMWs and Suzuki jeeps, I step out of the car a very secure man. It is clear I don't need my car to speak for me. It is said that a man with an expensive car is trying to compensate for a failing in some other department. In that situation the Maxi gives me status.

Different people get different status from the car. One man rang me and said that his workmates had thrust my *Guardian* advert in front of him, as his Maxi gave him the reputation of being the office eccentric. One person wrote that he was just like the Maxi, not very handsome but totally reliable. Why not? Dog owners are often said to resemble their pets after a while.

One caller told me that he was pretty sure Steve Davis drove a Maxi. I was thrilled, celebrity endorsement of the highest order. I rang Steve and to my delight he agreed to speak to me. Barry Hearn bought him an off-colour orange Maxi when he was an up and coming professional to get him around the country doing exhibitions, it was a case of have cue and travel, only in Steve's case it was in an Austin Maxi. He said, in true Steve Davis style, he had a few problems, but nothing spectacular. He eventually got rid of it when, with his winnings from his first UK Championship win, he bought a new Rover.

Later in his professional life, when the media all agreed Steve's image was that of a very boring man, he got together with the guy who did his *Spitting Image* character to put out the book *How To Be Really Interesting*. When it was discovered he used to drive a Maxi, they decided there was no better way to illustrate a boring man than to put him next to a Maxi. They found one, leaned Steve against it and took a picture. I've searched high and low for that photo, but no one I know has admitted to owning a copy.

As one picture says a thousand words, so does one Maxi. Could Basil Fawlty have driven any other car? When the National Lampoon team were looking for a car to symbolise the eccentricity of the British for their film *European Vacation*, they looked no further than a bright yellow Austin Maxi. The disregard for our culture by the Americans was demonstrated by Chevy Chase demolishing Stonehenge as he struggled with the gear change.

We are all a product of our environment. If there is something we cherish, it plays a major role in our life and will ultimately influence our beliefs but more immediately determine what kind of day we have.

The following stories illustrate the character of the Maxi and the status we acquire by getting in one.

Safari, So Goody

The Maxi seems to be the obvious car to paint by hand with a big emulsion brush.

The original Austin colours are far from inspiring. It is often more exciting to throw on your excess emulsion and watch it dry, than stay with the original colours.

The Maxi, The Frying Pan and George The Pig

To my mind the Austin Maxi was a paradigm of a seventies childhood. It was the car everyone's best friend's dad owned. It was also a favourite of the school teacher; usually English or History. In the eighties it became a firm favourite of the less ambitious self-employed painter and decorator - the back seats permanently down and a ladder bound with bungee chords to an ill-fitting roof rack. However in 1988, whilst employed as a laboratory technician, I purchased my first car; a nine year old Oxford Blue Austin Maxi.

My 'Blue Maxi' seemed the obvious choice. Kendall drove a caravanette, large but under the circumstances wholly unsuitable. I collected Kendall at 6.00 am prompt. We had a lot of work to do before the journey had even begun. I'd brought with me a large quantity of bin liners and the enormous pile of *Southampton Advertisers* from the kitchen which, in an ideal world, would have found their way to the recycling depot many months earlier. Kendall had done his homework too and reused a huge wooden crate from a skip outside Vosper Thorneycroft. A huge crate that nonetheless was to fit perfectly into the back of the Maxi.

We folded down the rear seats and ditched the rear shelf, which over the car's previous nine year life had buckled considerably under the sunlight and in fact had never fitted properly since the day I bought the car. I lined the back firstly with newspapers, binliners and again with newspapers, Finally we lifted the crate into the back. It was as if it had been tailor-made but this was true of any large item when it came to the Austin Maxi. I'd named the car the 'Blue Max' after the medal awarded to German fighter pilots during the First World War. On reflection I might easily have chosen to call it the Tardis but I wasn't a doctor and despite some of its incredible journeys, time travel, sadly, was not amongst them.

The conversion complete, and with the appropriate legal documentation in hand, we headed north out of the city along the A33. As we drove, the circumstances of our bizarre guest dominated the conversation. Our boss's sudden retirement had come as no surprise. Some were glad to see him go whilst Kendall and I were saddened by his departure. Not least because we were worried about our own respective futures in the department.

There was no question of a leaving do; the circumstances surrounding his departure were too clouded. Nonetheless, a modest twenty pounds had been collected from his minions with the intention of buying something to remember us by. It was the same twenty pounds that now sat on the dashboard accompanying the legal documentation which was shortly to be exchanged at 7.30 am at a crossroads in the middle of the Hampshire countryside. Our boss was head of our department which had seen him receive a good salary. So good, that he'd bought a dream cottage in the country with a huge garden which any aspiring country squire would be proud of. As he often reminded us, so huge was this garden and so far into the countryside that he didn't even need planning permission to build outhouses or even, for that matter, to keep livestock. Modesty wasn't his greatest asset and the constant emphasis on the size of his estate helped considerably in our choice of a leaving present. We'd decided to buy him a pig.

If he didn't want it - tough. He had the farm, he could afford to feed it and as leaving presents went, hopefully he would not forget this one in a hurry.

It seemed inherent in the nature of the farming community that if you wanted to buy something from a farmer then 7.30 am was as good a time as any. We'd worked hard on the 'Blue Max' - and in my humble, though limited opinion, the back of that car could have safely taken a pack of lions or a brace of velociraptors from Jurassic Park. We pulled up at the appointed crossroads and waited. At 7.30 precisely a pick-up truck pulled up next to us and out stepped Mr Pig Farmer.

After a brief exchange of pleasantries we gave him the cash and pig - moving certificate. We then gestured towards our magnificent mobile pig pen. The farmer was suitably impressed but felt compelled to tell us that the pig's current repository, a small cardboard box, was more than sufficient and that, if anything, a touch of overkill on our part had crept in. However, the large crate in the back of the Maxi was just the sort of thing that your resourceful farming type 'could make good use of '. So we cleared the case out of the back and tucked the little cardboard box with its piglet occupant into a corner and headed home. We were happy, the farmer was twenty pounds and a huge crate better off and the pig was happy. In fact, everyone was happy. Even our boss was happy as he lifted the cardboard box out of the car. But then anyone would have been, it was a fairly new box and it had a Sanyo logo on the side. The occupant of the box was accepted with little reluctance and became a much talked about family pet called George. And the car? Well the 'Blue Max' remained with me for another year. Sadly, five years on, I suspect that it has gone the way of most Maxis and was recycled into a set of frying pans or some such item. If this was the case then it is not beyond the realms of possibility that at some point its remains were briefly reunited with those of George the pig.

I was blessed, cursed, or simply unfortunate enough, to be given a 1750 Maxi to provide transport as an airline representative covering the East of Scotland. At times the job could be pretty boring so I would take interesting scenic routes to my given destination. In this instance I had to travel from Edinburgh to Inverness so I chose the quite dramatic route via Perth and Braemar on the A93, over the Lecht A939, the highest public road in Britain, to Nairn. Just short of Tomintoul I changed down to enter the village and found no gears at all. In my hand I had a pudding stick.

So I phoned the AA and eventually the familiar yellow Land Rover arrived; the Maxi was hoisted onto a sort of cradle and we set off back towards Ballater. At the junction of the A939 and the B973 a single track, very humpbacked bridge crosses the embryo River Don at a place called Colnabaichin. Sitting in the cab of the Land Rover, the bridge looked an awful lot narrower than I would have expected and I felt a little alarmed as the dolly to which the Maxi was attached creaked, groaned and rattled as we climbed the bridge. At the apex there was a loud bang followed by a metallic scraping noise. The driver muttered a few oaths and leapt out along the bridge parapet. 'It has come off the effin cradle and has jammed us on the bridge.' I climbed out to assist but out of the corner of my eye I noticed a convoy of olive green Range Rovers approaching somewhat cautiously. The leading vehicle had a large galvanised wire basket on the front loaded with dead pheasants. Suddenly, an explosion of sinister looking people; menacing, well dressed men dashing in all directions. The penny dropped! The leading Range Rover contained none other than HM the Queen and the rest of the family behind in other vehicles. This incident happened in 1975 at a time when the IRA started to become very hostile. The security forces obviously thought this was an ambush and my Maxi was the obvious target. It took a lot of explaining to sort out the problem but once all the heavies got stuck in and cleared the bridge the convoy continued with a smile and wave from the monarch.

Riverside Garage, Ballater, by appointment Motor Engineers to HM Queen, took the Maxi under their wing. It lay in that garage for seven weeks because they couldn't get the spare circlip to lock in the spur gear that had dropped off. I finally had a blazing row with them, found the part in Glasgow, posted it normal First Class to Ballater and had the car back in a few days.

This time, just outside Lecht, heading south on a very warm clear July afternoon, I was giving the Maxi a bit of downhill welly on the Braemar to Blairgowrie road. As I recall, when you got the Maxi wound up, the fan used to make a hell of a noise and I was doing just that southbound from Devils Elbow coming into a series of interesting but clear bends. I braked hard for the last one, turned the corner to enter ... yes! another single track bridge over a stream, when a Range Rover shot round the opposite corner going a fair old rate. The reflex action was...'Mmmmerd!!!' Both vehicles stopped in the centre of the bridge with the respective drivers' heads hanging out trying to assess if there was enough clearance to pass. I had the sun in my eyes and could only make out a mop of blond hair that looked rather sassy, a whiff of expensive perfume and a voice from that direction which called 'Am I alright?'

To an unwed macho as I was at that time, the answer had to be, 'Oooh yeah, you bet.' The Range Rover drew alongside. This time it wasn't the Queen but Princess Anne with Mark Phillips in the passenger seat.

HRH always has had the unique ability to bury you in a ten foot hole with a withering look. This incident took place well before their engagement and it is worthy of note that the Royal family was not in residence at Balmoral at the time.

Another abiding memory concerning the Maxi happened during the miners' strike in the 1970s. I always used to call at the grocer shop in Tomintoul for refreshments and I remember seeing small paraffin lamps on the shelf on a previous occasion.

Getting candles was difficult so I decided to lob in and buy a lamp. Returning from Inverness I headed to the grocers in Tomintoul, but by the time I got there the last lamp was sold. I pressed on towards Ballater and it began to snow. I was conscious of a Ford Escort Sport snapping at my heels making a lot of noise trying to pass on the single track section. By this time the road surface was getting very bad and the snow storm became a blizzard and I was climbing into it! Just where Banffshire borders Aberdeenshire the road rises to just over 2000ft and at this point the snow became too deep. I stopped, and the Escort with difficulty managed to pass. With disgust I watched it howl into the distance. I knew that another few hundred yards would have taken me over the summit and down into the shelter of the Dee valley. I sat there watching the conditions get worse until I could see nothing at all. A complete white out. The engine was still running but I had to get out and clear the drifting from the front of the car and the radiator. Quick calculations suggested more than enough fuel to sit it out for a long time. But how long? Fear took over and I was asking myself what the hell I was doing on this road at this time of year.

Quite suddenly a Land Rover appeared and torches began to hop around the scenery. People! Help! The story that unfolded was more than welcome. The Escort belonged to the son of the pub licensee at Corgarf further down the road. The lad had told his father about the Maxi so because it hadn't passed through, they set off to rescue me.

Sitting in the kitchen of the pub later, the landlord asked me what I was doing in that neck of the woods so I drew his attention to the strike and the paraffin lamps. He threw an old-fashioned look, climbed on a stool, reached above the large fridge freezer and tossed a large box of candles in my direction. 'Here! Take these, we don't need them.' When I asked him why not, he told me they generated their own power so the strike didn't bother them at all. Then he told me that a couple of weeks before someone had been found dead from exposure seated in a car at the same location. Also, the driver of the Escort had stopped and filled his boot with rocks. Want to debate the merits of front wheel drive?

I am now 24 years old, and ever since I can remember, our family has been the proud owner of a Maxi. When my sister and I were old enough to learn to drive, a second Maxi was bought. And, at one time, we were in proud possession of a 'fleet' of three Maxis. The fact that the body of the light blue one, whose engine was being used to replace that of the cream one, was only fit for the breaker's yard is by the by. My mother had been very surprised when the engine of our 'faithful' cream Maxi had suddenly seized on the motorway; my father was somewhat less surprised upon discovering that she had been driving it with no oil or water. Cream ones, white ones, orange ones, blue ones and the inevitable shit-brown ones, we had them all. A turning point in our Maxi ownership came when we upgraded from ones with shiny, plastic seats to ones with 'deluxe' material-covered seats. Such luxury!

When telling friends that I drove a Maxi, the standard answer used to be, 'Oh yes, our family had one of those ONCE.' The Maxi is indeed the epitome of the family car, being large, roomy and practical. It was probably also the inspiration for the development of transformers. For the Maxi is not simply a car. It doubles up quite successfully as a home, a

Snap

If you were driving past a billboard and saw a picture of a car that looked like yours, would you pull over, climb out and take a picture of it? Chances are you wouldn't, but if you were driving a Maxi you would at least think about it.

Maxi drivers are seldom celebrated by the advertising image makers, we have to capture the moment.

bedroom or a removal van. As children on a family camping holiday in France, we arrived late at night at one campsite to find that the barriers were already down and that entry was to be refused. Not to be deterred by French bloody-mindedness, we simply parked our Maxi on the verge by the barriers, put the back-seats down and used it as a caravan. My mother, my sister, my brother and myself spent a very comfortable night in the Maximobile. My father chose to revert to nature and sleep half under the stars, half under the car! He was found there the next morning by the campsite owner who had the nerve to demand payment from him for our pitch on the verge. The Maxi was quick to revert to a car and to zoom off at top speed, a gallant 20 mph.

My sister and I had reached the age when we felt pride in our parents' choice of car. My brother had however reached that sensitive age at school when it was only deemed cool if you could roll up at school in an XR3pqyz, or some such monstrosity, preferably with turbo and injection to add to the street cred. My father was often heard to puzzle over the fact that my brother always asked to be dropped off about half a mile away from the school because 'he enjoyed a short walk before being cooped up in a classroom all day.' I have a sneaky feeling that the acquisition of a turbo injection something-or-other might well have curbed his desire for fresh air!

I was always pleasantly surprised to find a steaming cup of tea waiting for me when I turned up unexpected at friends' houses. I used to think that I had a very large number of telepathic friends until one of them let slip that they were always prepared for my arrival because they could hear the Maxi clanking away at the end of the street well in advance! This advance-warning feature of the Maxi was not so favourable when I arrived home late at night, not wanting my parents to realise just quite how late/early it really was! Our Maxis have become our mascot and, rather hurtfully, a source of constant amusement amongst our friends. Our surname is Clark. When the film, *National Lampoons European Vacation* showed the American father, Clark, driving a Maxi none-too-successfully around London, it caused uproar amongst our friends.

O ur first Maxi was the original model with the awkward gear change which we got used to, but the second was a real winner. Although it was only a 1500, the engine was really smooth and the car was superior to a friend's 1750. We have two sons, both young then, and went everywhere in the Maxi. It was also used for work - ferrying people around. I was a probation officer then, travelling all over the country, and it never let me down. Three situations come particularly to mind when remembering the Maxi. On one occasion, a brilliant summer's day, I was driving my sons, then aged seven and four, to visit their grandmother. We were travelling cross country from Hereford to Ashampstead on the Berkshire Downs, and as we drove through Wantage, a young man in a new Morgan passed us. We followed him over the B4494 downs road to Farnborough. The road is narrow, hilly, with one bend after another. The sports car accelerated away or tried to, but I followed at a steady 50 and it became apparent that on the bends he was scrabbling round while we were taking them smoothly. I was not attempting to compete, but it was so clear that with our wider track and front wheel drive, the Maxi was just better suited to the road than the expensive sports car. A remarkable experience for me and my sons, who henceforth regarded me as a great driver. Not true! It was the Maxi.

About once a month we drove to my parents to visit on a Sunday, and, at the end of the day, put the boys into their pyjamas, folded down that unique back seat, tucked them in with a blanket, and they slept comfortably all the way home - about 120 miles. On arrival, we'd carry them, still asleep, upstairs, and pop them into bed. Magic! What other car could you do that with?

Finally, a winter story, which illustrates the roadholding and balance of the car. We left Cirencester in heavy snow, taking the Roman road, which crosses the Cotswold ridge and eventually drops down to Gloucester. Conditions were terrible, the snow was a foot deep, and there were times when we lost adhesion, but careful use of the throttle kept us going. Then we reached the top of Birdlip Hill just outside Gloucester, the police were manning a barrier and stopped us to ask where we had come from. When I told them, they said the road was considered impassable and was now closed. We apparently got through just before they had closed the road outside Cirencester. We had also, incidentally, picked up a driver on the way who had abandoned his lorry. The police were impressed, as we all were, with the Maxi's sterling performance. What a car! Happy memories.

I have a friend called Rob, who is now in Australia, but I first knew him from about 1980. He was a bit of an amateur car mechanic - but his skills only extended as far as getting an old wreck going for his own use. Around about 1981 he had two Maxis in succession. The characteristic that distinguished them from all the other cars he had had was that the batteries for both resided on the front passenger seat. Apparently he could never get the right battery to fit in the engine, and so they had to take pride of place, with wires curling through the windows and dreadful eaten-away patches from splashed acid. We both lived in Kent at the time, near Sevenoaks, and various groups of friends would tear around the country lanes in these cars, going from pub to pub, and stopping to collect extra passengers.

I remember the Maxis as being so spacious and almost stately, and even though they were incongruous with our images of ourselves - being eighteen at the time, a couple of friends drove Capris - they had a certain elegance and style. One chief defect of one of the Maxis was the tendency for the back seat to collapse when under pressure, and on at least one occasion I was in a car behind and was astonished to suddenly notice that the heads of the three people on the back seat had mysteriously disappeared. And being roomy, these cars had other uses; more than once, while waiting for a lift home from Rob, I was embarrassed to realise that the suspension of the parked Maxi was being strenuously tested, and the windows had steamed up...

When I moved down to Devon from London last November, I asked the local taxi driver - who also deals in second hand cars - if he could find me an old banger which would get me to and from the station, about eight miles in all. He sold me a Maxi for £325, and since then, I have spent £250 on it on repairs. It refused to start when I first got it so I called the garage out and they spent an hour charging the battery and checking the plugs and points and then they discovered it hadn't got any petrol in it. It has had two bad oil leaks, one water leak, two new tyres, and a dicky alternator.

In spite of all this, I have got quite fond of him in a funny sort of way. I called him Malcolm and gradually got used to his little eccentricities. Two people are needed to open the boot - one to put the back seat down and crawl inside and hold two bits of wire together, and the other to turn the key in the boot lock and pull up the lid. Then the inside person has to pass the broom handle out to the outside person who then props the lid open. Once the boot is open and the back seat down, Malcolm will hold an amazing amount of stuff. I am moving house soon, and apart from my grand piano which is going in a van, all the rest of my stuff is going to be transported in Malcolm.

The navy-blue, brown interiored OBM 560M 1750, came to me courtesy of my grandfather for the sum of £1. It was his second; his first bought new in 1972, lasted 180,000 miles. WNM 4K will always be part of my childhood folklore. At the age of eight, with Grandmother in the front, my brother and I in our enormous back seat went on countless expeditions, swimming, picnicking, and holidays. The day it turned up it was scrutinised by me quite thoroughly and its character is part of the house I grew up in. He kept it until 1981, and then passed it on, on its second engine.

His Maxi ownership continued with what later became a notorious landmark, and distraction for grazing horses. In the meantime, it was my first big car.

In 1986, at the age of 76, he decided to give up motoring. The car was as welcome as our first child. My then partner was 6 months pregnant. We also had a huge and beautiful Lurcher Dog, who was very happy in his part-time home in the boot. Recent journeys in my Morris Minor van in midwinter were proving cold, bumpy, and as unwelcome to a heavily pregnant woman as morning sickness.

So the affair with 'Blue Max' began well. And went on to some notable episodes, a trip to Denmark in -24°C where it started first time every time. The same trip, when offered a VW Beetle engine at a too-good-to-miss price, I hack-sawed the front seat out to create space to bring it home. The many markets where we'd put up stall, loading it to within inches of the road surface. The trip to see a reggae superstar with a carload of intoxicated amigos, when the weight in the back tested the front-wheel drive and spongy suspension to the limit, causing many walls to flash before our eyes. Perhaps the awkward flat steering wheel, the ancient heater-ashtray arrangement on a peculiar, veneered dashboard, its own peculiarities of a leak half way up its petrol tank, and that odour of melting vinyl one encounters when opening the doors on a sunny afternoon. These things would mean nothing but for the sentimental throwback to my childhood, and would have long since been confined to the past.

But of all experiences the trip back from Ireland was the most memorable in the car's time with me. Baby born, and our house in disarray, with our thoughts on the future, and spurred on by friends disappearing across St. George's Channel, we plotted to emigrate. This one of many questionable decisions made in a doomed relationship. The vehicle, loaded with bicycle and two trunks on the roof, dog in basket in the boot, mother, child, suitcases and baskets of baby stuff on the backseat, and yet more odd gubbins in the passenger seat, plus tape player running from the cigarette lighter socket in the front.

With this combination, the 5-speed hulk hit the Emerald Isle, my fatherland, a wet miserable West Cork, on June 21st 1986. The intention to settle and plant Curly Kale dwindled with our resources, and as they trickled away, it turned into a holiday without a home to return to. We blew it. And facing circumstances so trying, we planned a tour up the west coast before high-tailing to the nightboat at Rosslaire.

The car had taken all day stints under heavy burden as ever without a murmur of discontent from its growling engine. The engine never proved to be the car's downfall. It was very much our 'camel in the desert' providing four berth shelter at night. With the front seats pushed as far forward as they'd go, and fully reclined, the three humans were comfortable, and behind the rear seat's back a contented mumbling from the hound. So a camel by day and a Bedouin by night. Back in Britain, the journey to Dorset, broken by a short nap in Bristol, was entering its last lap. The sun was shining, which lifted our spirits, despite the gloom of defeat, and chatting of our disaster, above the delicate fragrance of nappies, we rounded a bend at the Southern boundary of the Somerset village of Compton Dunden, and began climbing a steep hill. I changed down from fifth, and engaged the clutch to attempt third. Nothing, except a shrill metallic whine. Again I tried, no luck, the car lost speed while coasting up the hill round a blind bend ..... and stopped.

No gear was available. Hazards on, I opened the hood, took a look, and after a brief inspection, found an empty clutch master cylinder reservoir, and a trail of fluid from the slave cylinder. Aha, not a new gearbox as first feared. Then I walked to the side of the car to report the good/bad news. Whereupon a number of things happened in rapid succession. I had scarcely spoken, when I heard the approaching roar, my instincts propelled me to the front of the car. A red Cavalier, overtaking a 7 ton Bedford T.K. Bulk Tanker, was flying round the bend, and into the path of a Fiat 126 coming down the hill, and a stationary Maxi, with a worried looking canine popping his head up at the back window, a young woman with a three month old baby in the back seat, and a ginger haired hippy leaning into the back window.

There was a very loud shrieking of brakes, horns, and, most audible, the sound of a 7 ton lorry striking a tree at between 40 and 50mph and practically disintegrating before my eyes.

Why that man elected to haul his wagon into the left bank 15ft from the rear of the clutchless heap, and why the tree, the only plant over-hanging the road, happened to be there at that moment and consequently absorb the massive impact, could probably only be assessed in terms of luck or religion The other oddity of the event was the driver of the Fiat being an off-duty police sergeant, who instantly began directing traffic around the site, and provided a credible alibi for the trucker. The Cavalier driver crawled stealthily away up the hill, checking his rear view mirror all the while, then sped off. The driver of the truck stumbled out and muttered obscenities.

I, having now run forward toward the wreckage, made the assessment that, having had his brush with death, the truck driver was now about to inflict a similar experience on my quivering frame. Mercifully the gallant geezer, who had just saved us by his unselfish action, was aiming his vitriol up the hill at the 'Fleeing Cavalier'. As well it might be, in fact anywhere but me please.

Very fortunately, no bones broken, the afternoon wore on, no available culprit and with my ceaseless thank-yous, the arrival of the traffic police, and the retired engineer living nearby who fashioned a clutch-seal from an old piece of rubber, to give us 'four changes home'. We got back the last thirty miles gently in third gear. The following day I purchased a slave-cylinder repair kit, price £1.20.

I found myself on that stretch of road about a year ago, heading in the opposite direction, and saw the stump of the tree, apparently felled by the council's chainsaw. It still bears a vertical scar, the white and green paint now weathered away.

The car, which I sold late in the following year as it approached its 100,000, was to make one more dramatic final journey itself. For £25 I sold the MOT-less 'Blue Max' to a licenceless friend who is a traveller, he towed his caravan from one site to the next, then sold it on again for £30 to someone who intended to take it into a local town and 'learn to drive in a car park'. They weren't to know at the time, but that particular Maxi was destined only for another six hundred yards.

He got in the front seat and put his girlfriend and baby in the back, then backed ferociously up a pile of gravel and got the car completely stuck. His girlfriend and baby emerged, she clutching her neck, cursing him, the baby screaming. She was persuaded to take an alternative lift while the lads helped dig the car out. So after much manoeuvring, the car freed, he sped off out of the lay-by, and onto a main road, and was instantly followed, lights a-blazing, by a police car, who'd been sitting at the entrance awaiting just such an event. Panicking he roared into a side road down a hill to a T-junction and went straight over it, through a hedge and into a field. The car stayed in this place in the corner of the field, and was clearly visible from the A35 trunk road near Bridport for the next three years.

Sometimes as I drove by, a horse would be rubbing his neck on the angled boot. One winter I noticed the farmer was using it to store hay. Everytime I saw it I'd get a touching memory: the holiday, the accident, my grandfather mainly, who had since died. I don't know what he'd have made of his car coming to such an end, it certainly wasn't on the cards when I drove away in my £1 hatchback from hell. One day there was a bare patch of grass in the corner of the field, and a cube of bluey-brown metal in the local scrap yard.

I used to live in South Africa where the Maxi was neither assembled nor imported but I do remember reading about it in a motoring magazine and thinking, 'What a sensible car'. I emigrated to England in 1971 and had this impression confirmed. South Africa had the 1100 and the very stretched 1800 or 'land crab' but not the size in-between which I think was a silly omission.

In the 70s I developed a irritating ear complaint which made it very unpleasant for me to drive in any car in which a certain type of resonance occurred. I would get earache, migraine, dizziness - you name it, I got it. I had to try out various cars to find whether they suited, and the awkward thing was that it only showed up after about 3 weeks or 300 miles. As you can imagine it was pretty expensive. Over a few years I bought and quickly had to sell a Triumph Dolomite, a Renault 12, Volvo 343, a Citroen BX, a Triumph Acclaim and some others. The most nearly satisfactory were the 1100 and the 1800. But finally I found the Maxi and, provided I did not remove the parcel shelf at the back, it was OK - but deadly with the shelf removed!

My first was a 1972 1500 which, though absurdly underpowered, was nice and quiet - at least as far as resonance was concerned. I became very fond of this slow, lumbering old thing. Having had an Alpha Romeo Giulia in South Africa I found the Maxi hopelessly unresponsive to drive but what a wonderful load-carrier. I also managed to lay down my six foot frame in it a few times and slept in considerable comfort, but my wife, alas, would not join me!

I bought a new Maxi in about 1986, which kept me going until 1991. Over this period my ear trouble had worsened to the extent that I couldn't travel in any car other than the Maxi and this created a close bond between us. However, the rust set in and though I patched and filled, MOTs were becoming tougher to pass. My search had to start over again and most Maxis were now well past their sell-by date.

My friends were always very rude about Maxis but I know they envied the lorry-licks load-carrying capacity. What's more they were both very reliable. Ironically the only unreasonable expense I had was when I hit a bad pothole and damaged the front suspension hydroelastic unit. I had to pay £250 to have it fixed a few months before rust caused it to fail an MOT and relegated it to the scrapyard. That was a sad day.

My General Sales Manager had done a 'deal' to purchase a handful of 1750s, so I was summoned down to London somewhere in the Epsom area to collect mine. Along with the Newcastle rep we arrived at a large basement car park to be faced with two brand-new vehicles, one maroon and the other white. Quick as a flash my colleague snapped, 'The white one's mine!' And so I was stuck with t'other. Not my favourite colour! A company vehicle is always a bone of contention with reps. It is a status symbol. If it's a GT or a something 'i'...you're somebody. But a 1750 Maxi. We were off to a bad start. A white Maxi perhaps redressed the balance a little, but the other one? No way! To complicate things even more, because it was a foreign company they had a peculiar attitude to running business transport. We had to keep them for seven years or 100,000 miles, whichever came first. Try keeping up with the Joneses with that formula.

We had been given strict instructions to stick to 50 mph all the way home and up to the first service at 500 miles. Reasonable enough! But at that time I had just read Lady Antonia Fraser's excellent book *Mary Queen of Scots*. I decided to call at Fotheringay just off the A1 on the way north and see what was left of the castle where the great queen was executed. Eventually I found the site at the back of a farm yard. As stated in the book the castle was pulled down to prevent it becoming a sort of shrine. It is still an interesting place to visit. What had been a moat is still visible, with a steep hillock behind it on top of which the castle had once stood. I had great difficulty climbing up the slippery bank which was helped by grasping branches of thistles. Once on top the outline of the walls is very evident, also the location of the Great Hall where the execution took place. I stood for a while taking in the beautiful view and tried to imagine how it appeared to Mary Queen of Scots who had been imprisoned for sixteen years. The romantic side of my Scottish heritage took. In the place where I assumed she met her end stood an enormous clump of thistle, the biggest I had ever seen. Attached to my key ring was a small pen knife, one of those airline giveaways that you don't see so much these days. I tried to cut free a stalk of thistle with a very large flower and it proved extremely difficult, almost impossible! Was this an omen? A reminder of the queen who never should have been beheaded? I returned to the Maxi carrying the sprig of thistle and headed north.

By the time I had levelled with Doncaster en route to Edinburgh I was suffering a toothache in my backside and leg pain from the atrocious driving position. I decided to call on my chum at Grimsby and spend a couple of nights there. Now Bernard is one of those exceptionally talented fellows who can disappear into a very limited toolbox and perform miracles. By the end of Saturday we had revamped the Maxi driver's throne using an assortment of mini brackets and repositioning anchor points. The seat was given a couple of extra inches rearward adjustment and the steering column lowered to put everything in the correct place. Later I made a special seat cover padded with foam to keep my bum attached to the car. This made the Maxi much more comfortable to drive over long distances; if I remember correctly first gear was almost the only out of reach bit; that was a small price to pay for overall comfort.

The drive from Grimsby to Edinburgh was boring and uneventful. However, I had time to think about the sprig of thistle and decided on arrival to take it to Hollyrood Palace and toss it over the garden wall. No one has ever mentioned an outbreak of thistles in the palace garden but if there has been, then Mary Queen of Scots is back in residence and they are going to have a hell of a job getting rid of her next time.

Steering the Maxi could be compared to manoeuvring the 'Queen Mary' with the quayside still roped to the ship. I tried every combination of tyre pressures and even tyre brands to improve handling but to no avail.

It took 87,000 miles before the Maxi developed suede bumpers and extensive clusters of rust bubbles about the bodywork and I had to get rid of it. Mechanically, it wasn't worth spending any more money on. It got to the state I flatly refused to drive it! Eventually the company decided to take its revenge for my ingratitude by allocating a new Datsun 140J Violet. 'A Violet!' I screamed at the salesman who handed me the keys. 'What kind of name is that to give a motor car?' Out of the frying pan into the fire, for another seven years! Anyhow, it was company practice to send a memo to all employees asking if they would care to make an offer for discarded vehicles. I received a telephone call from a London colleague who questioned me about the Maxi's condition. He wanted it for a second car, a runabout for his wife. I replied it would be more suitable for his mother-in-law, it really wasn't a good buy. His silly offer was accepted and he became the new owner of WMV, the only three digits of the registration I can remember.

I heard that it only made one short journey after I sold it. As five heavy men tried to squeeze in, there was a big crash and they all ended up on the tarmac. The floor ripped out under the strain. It ended its days sitting proudly atop a big metal heap somewhere in an East London scrapyard.

I had received from Head Office a letter asking me to meet one of the directors of the company. Already very short of time because of my forthcoming wedding preparations, I gazed with dismay at the badly damaged driver's door of my Maxi. It just had to be fixed, but how? Someone suggested a wee man called George who worked from a former fish house down at the water's edge behind Preston Pans. George located another door at a scrapyard, fitted the door and realising the tartan tank image wouldn't do, he sprayed it the correct shade of maroon and swapped the trim. I was relieved and very grateful. Looking the thing over George disappeared into the pile of bent wings and car bits to emerge clutching a violin in his blackened hands and started to play a selection of snips from heavy classical pieces followed by a couple of Scottish reels. I was gobsmacked but able to ask, 'How much for the job?' His right index finger stabbed his greasy cap about a half inch above its normal plimsol line and he thoughtfully replied, 'Nae a big job laddie, sixteen quid for the door, a fiver for the paint and a couple of quid for the music... say twenty five notes'. I didn't want to query his arithmetic.

About eight months after I bought my Maxi, a Triumph Spitfire took off as it negotiated a hump-backed bridge whilst coming towards me. Owing to a slight curve in the road, he landed partly by raking my front offside wing.

Since the address he gave me turned out to be a Public Library, and his plates were traced to a fire engine in Aberdeen, I was unable to claim, as my insurance was only third party fire and theft. I beat the wing out with a mallet and drove happily for the rest of my ownership revelling in the corrugated look.

One of the main weaknesses of the car, as far as I was concerned, lay in the tubular construction of the front seats. Those tended to stress-crack near the adjustor and I believe many Maxi owners had welding done, or replacement seats fitted.

I had had an early motor cruise control fitted, and this locked the throttle at the setting required. Having set the car thus, my cousin and I were chatting away when there was a loud 'crack', and I found myself deposited on the back seat! My cousin found himself in a driverless car doing 70mph in the fast lane on a very busy motorway. The driver's seat had collapsed. I eventually managed to get back behind the wheel, but there was a very unusual aroma in the car for a few days afterwards

Priding myself on my car's abilities, I once accepted a bet that I could go from Llangellen Highstreet to the cafe at the top of the Horseshoe Pass without dropping out of fifth gear. There are several hairpins and the gradient goes to 1 in 10 or worse. Still, I did it and won my £1.00 bet. There were some very ominous noises coming from the engine and transmission, particularly as the car was carrying three hefty adults in the attempt, but the car never actually stalled.

Having got to the top and stopped for refreshments, I noticed a hill climb competition going on, on the high ground behind the cafe car park. Being still rather full of myself following my recent success, I enquired of the Marshalls, and found that entry was open, at 50p per car. I paid up and came third, winning £5.00. In fact, only lack of ground clearance stopped me from winning and this was against specialist hill climb vehicles. Certain competitors, mostly from fourth place down, were not best pleased, particularly when I assured them that it was a bog-standard production car, apart from Firestone Town or Country tyres on the front.

Another time, my mum, dad, nan and I set of to go to Newquay in Wales for the day. All went well, but as we entered Carmarthen, I pointed out of the windscreen at a wheel bouncing down the road, saying 'Hey Dad, some fool's lost his wheel!' Those words were swiftly followed by a sickening lurch of the car and a horrendous grinding noise as I stopped the car using the bottom of the nearside front wing instead of brakes.

It was my front nearside wheel, together with brake disc and part of the drive that we had soon merrily bouncing away; in the words of the poet, something had snapped.

We came unceremoniously to rest in the middle of the driveway to the Bishop of Carmarthen's Palace. It was the christening of his grandchild, and, kind man that he was, we were invited to join the garden party whilst waiting for the AA. We've still never been to Newquay.

Practical

It may not be big enough to get lost in,
The Austin Maxi.
But it's bigger enough to make a double bed,
Because the front seats go right back see,
In the Maxi.

In my dim and distant youth, I only had one piece of information on the Maxi. It came to the fore later in life when I was looking for my first car and spotted a Maxi in the classified section of the *Express and Star*, Wolverhampton's daily newspaper. I didn't know what one looked like, I had no idea of their reputation, I certainly had no idea what I was letting myself in for, but I did have this small piece of information that was passed on to me by my dad when I asked him,

'Dad, what are Maxis like?'

'Maxi stands for maximum interior space for minimum outside body, lad,' he told me with the authority he always assumed when he wasn't entirely sure, but wanted to give an informed reply to my persistent questions.

I took this scrap of information and fitted it into the equation of the classified ad - 'Austin Maxi 1750, twelve months MOT, £195' - and my self-imposed rule that I would buy a car that was strong, could get me long distances without complaining, that could fit my bags in, that was in good condition and that was cheap. It fitted the bill enough for me to go look at it and as soon as I saw it parked outside an old man's house, with its bent front grill giving it the look of a sad face, I had found my first car.

So the initial selling point to me of the Maxi was a solid, no-nonsense motorcar. The character and status and humour of it all came to me later, but all the while pumping good time behind all of its absurdities was the solid back-beat of strength, space, power, comfort and quality. It can tow a caravan, it can fit your family and dog in the back, if you want to throw a washing machine in the back there is no problem and if you are ever stuck for a place to lay your head, you can always fold the seats down to make a double bed. All this in a car that's not too big to park.

I bought my first Maxi. It took less than five miles for it to let me down drastically. The reason I wanted a car was because I was leaving college and being mobile was a big help in getting a job. I picked it up the day I was leaving. My housemate Adam had borrowed a college minibus to take all his belongings back to his dad's house on the Wirral. He was soon to be departing on a long trip round Australia looking up his long lost ageing relatives and building himself into their wills, so I wouldn't be seeing him for a while. It was a good cue for a party - at his dad's house. I was on my way back to Yorkshire and the Wirral was sort of on the way, so I decided I'd go in the Maxi and he could go in the minibus. I was also itching to get it on the open road and see what it went like.

We threw our bags in the back of our respective beasts and beat a track for 'out of town'. Driving your own car for the first time brings with it a certain buzz, especially when you have just put behind you your college years and are looking for a new start in life. I had invested my last £200 wisely and I was off into a brave new world.

On the outskirts of Wolverhampton is Tettenhall Hill. We approached it in convoy. Adam was in front of me and as we started to climb the hill I saw him getting away from me, I thought he was initiating a race so I went for the throttle only to find that I didn't have any, the Maxi was slowing down, big style. I changed down to third and kicked in again, nothing. I was down to second, the same. I was out of the saddle and whipping, I was in first gear and the Maxi shuddered to a halt. It wasn't going up that hill.

Now I'd made a big deal about buying this Maxi. My friends thought I was crazy, but I stood up against them and when I gave my farewell speech, it went along the lines of, 'Hey, be careful out there, because here comes Selwyn and he's not alone, he's got a Maxi.'

When Adam finally returned down the hill to find out where I was he couldn't have been smugger. I loaded my bags into the minibus and without ceremony, left the Maxi by the side of the road. I would go to the party and pick it up on the way back. It wasn't a good start to a new life.

When I returned to the car and got a mechanic to look at it, it transpired the problem was nothing more than a blocked-up air filter. He took it out and didn't bother putting another back in. The Maxi was flying and so was I.

My man at the *Guardian*, John Salt, was driving a Mini at that time and it had broken down in Wolverhampton town centre. He needed a tow, and when he saw me pull up in my big Maxi with a nice shiny towbar attached to the back, I was asked the obvious question. I was a bit nervous considering only a few days previous it wouldn't go up a hill, but he needed help so I agreed. We got out the rope, tied it up, his passenger got in with me and we set off on the five mile journey back to his house.

Salty's passenger was a friend of mine too, so we got into talking, having a laugh and forgetting about why we were in the car together. I was telling him all about the Maxi, putting my foot down to demonstrate its power and flying round corners to show him the road handling. It was only after one of my rare glances in the rear-view mirror that I suddenly remembered Salty in the Mini. Another glance saw a petrified, pale-faced man, grabbing onto a tiny steering-wheel for his life. He certainly knew of the strength of the Maxi and now, at last, so did I. It had finally justified my faith in buying it, I had made the right decision and my life wasn't going to be the complete disaster I had anticipated. When I got Salty home, I didn't get out and apologise, I got out and thanked him.

I have had many ups and downs with the Maxi since then, but I keep coming back to it because the bottom line is, it is a very sturdy car. Looking back on the original sales material it seems that its practical features were what it was sold upon. It was the first hatchback, it was the first five geared car. Someone wrote to me saying that they had heard that when the Maxi was first launched the marketing people at Austin had decided that there could be no better endorsement for the car than for Sterling Moss to drive one, so they sent him a brand new Maxi. After only a few weeks they were very disappointed to get it back with a little note from Sterling that said,

'Five doors, five gears, five effin miles per hour.'

That would have touched on a raw nerve. The original Maxis had the cable gear change, which after only a few miles motoring would stretch, making it impossible to find the gears. The motoring press had a field day with this and the Maxi suffered terribly. The Ford Cortina was around at the time and the British public lapped it up. It was hard times for the Maxi sales people. It was estimated that 150,000 Maxis would be sold each year and production was set to cater for this. With the knowledge that only 450,000 were sold in the whole twelve years of production, it is obvious that something went terribly wrong. I have heard that Maxis were stockpiled in aircraft hangers which were full of birds. The Maxis were apparently covered in the white stuff that falls out of a bird's backside. What a missed photo opportunity that was.

I did get a telephone call from a taxidermist from Cheshire who told me that he once got an order, from a zoo on the south coast, to stuff two lions. He drove down in the Maxi, folded the seats back, lifted the pair of lions into the back and drove them up the country back to Cheshire, the front wheels hardly touching the ground the whole journey. He added that he once stuffed a grizzly bear and was delivering it to its final resting place in a snowstorm. A fully stuffed grizzly was too big to fit in the Maxi without leaving the tailgate open. He said it was good fun driving past schools and you don't get much hassle from other motorists, when you have a snow covered bear hanging out the back of your car.

In terms of design, the Maxi couldn't have higher credentials. Alexander Isigonis was famed for cramming as much motor car into as tight a space as possible, making his name, fame and fortune on the Mini. The Maxi was a new challenge, it was to take all the benefits of the Mini and fit them into a bigger family saloon. He managed it, but didn't receive anywhere near the same acclaim.

I read the Isigonis orbituary in the *Guardian*, it only mentioned the Maxi in passing. It seemed as though the writer didn't want to stain his memory so soon after his death. The following section is full of stories by people who have stuck by their Maxis and have reaped the benefits of the design team's work, doing the kind of things that Isigonis hoped they would.

So Buy One

The sales people were keen to highlight the practical benefits of owning a Maxi. They were also keen to place them in prominent positions. The first Wimbledon tennis escort cars were a fleet of green Maxis that ferried players too and from the complex.

THE CAR FOR PEOPLE WITH MORE SENSE THAN MONEY

On a journey to France we were queuing for the ferry behind an Austin Maxi towing a small trailer. As the queue progressed the Maxi approached the ramp to the ferry. All the passengers and driver got out, they folded down the back seat and lifted the trailer into the back of the Maxi. The driver drove the Maxi onto the ferry with the others boarding as foot passengers. I thought this was rather an ingenious method of saving ferry costs as foot passengers went considerably cheaper and the overall length of the vehicle was only that of a Maxi instead of a Maxi plus trailer. I admired the forethought and design abilities of the driver in building a trailer which fitted within the Maxi.

I worked in the motor industry for 40 years in what is now described as the Rover Group before retiring early two years ago, working during that period for the Nuffield Organisation, Pressed Steel Co, British Motor Corporation, Leyland Cars, Pressed Steel Fisher and, after the final name change, the Rover Group. So many mergers and name changes happened in that period that I've probably missed a couple out! I commenced as a Nuffield apprentice and during my 40 years was a senior manager for 25 years mainly in manufacturing engineering.

My interest was kindled by the fact that I had at least five new Maxis and enjoyed them. My last was a red 1750, registration CUD 662R and I often wonder whether it's still around. My other great interest is that the Pressed Steel Co were the leading body designers, tooling manufacturers and car body manufacturers in the British car industry and we carried out all that work on the Maxi. I planned and controlled the whole programme from styling, through body engineering, model manufacturing, the planning and the design, patternshop and foundry work, jig planning and design, tool and jig manufacturing, tool and jig try-out, body pre-production and pilot build for the first thirty bodies built before handover to production. This was a time cycle of 32 months and involved about 1,400,000 man hours with enormous material costs. The car body cycle is the longest lead time in the production of a new car unless a brand new engine is involved and therefore these activities are under great pressure from start to finish.

This was Britain's most underrated car. My personal feeling was that the publicity information produced on this car never ever was successful, in that the customer never realised from it what was contained in the whole car package. I used to realise this first hand at the Motor Show at Earls Court, where Pressed Steel always had its own stand in addition to all the other marques of Austin, Morris and the others. I always did a stint, and on our stand we used to tell visitors that 'we're not salesmen, we design and make this car and we would like to tell you about that.' Visitors were very impressed and interested and our stand was very popular and always crowded. My colleague Egan Wiggins and I had a routine on the Maxi worked out on the stand which we used to perform over and over again. We would stand either side of the vehicle, open rear and front doors, tailgate and bonnet, adjust the rear and front seats down into their reclining positions and jump back into our starting positions with a loud 'Olé.' This was all done in a fraction of time with a bit of work study and practice, the crowd were simply amazed and used to flock round the car. On further examination of the inside and various features, ninety percent of them used to say, 'We had no idea that the car had all these features.' Many of them used to come back later and say, 'I spoke to you earlier on, I've been all round the show and there's nothing to compare with it.' We also used to discuss with people things like safety features in the body design and the torsional strength of the body which was far superior to all its competitors; again never mentioned in its publicity. Although strictly speaking it never had a straight competitor, there was nothing quite like it. We reported the knowledge gap regularly to the sales staff.

The car never had any serious problems although it was severely knocked by the press over a problem of cable stretch on the gearshift connections to the gear box, which was cured by changing to pre-stretched cables. I never experienced the problem myself. At this time, and for years afterwards, the knocking sections of the British press had field days on the British car industry, and whilst some criticism, particularily in the conduct of industrial relations, was justified, their attitude was astonishing compared to continental loyalties to domestic industry. This contributed greatly to the industry's decline. I found the cars I had superb, reasonably lively, economical, comfortable and they could easily accommodate three children in safety seats, a fortnight's holiday luggage and a dog. I never had a fault on any of them, what more could one want?

I well remember my last Maxi in 1978 when I moved house and designed a very large complicated garden and visited the garden centre at

Blenheim Palace. I bought several hundred pounds worth of large trees, shrubs and plants. I unloaded all the barrows and stood them in a vast array on the road at the rear of the centre and went and got the car. The manager, noticing the huge purchase, stood somewhat bemused. I put all the seats down and stacked the whole lot in the car. He was flabbergasted and said, 'I thought you would need a lorry for that lot.' I knew what he was inevitably going to say next. 'I never knew you could do all that with a Maxi.' Notice again, after years of production, the public just didn't know what the car was all about.

I was also very thankful for the safety features designed into the car body by my colleague Peter Finch of Pressed Steel who was one of the industry's most brilliant and progressive safety engineers. The front end of the car was designed to crumple in on impact between the right and left front wings to absorb impact forces. I was driving one day to work on a small country road early in the morning in half light. The road at that point went up a short steep slope, with no view of the bottom of the dip at the end of the other steep short slope down the other side. On reaching the peak of the slope up and going down the other side, I realised there was a car abandoned at the bottom of the dip. As I began to pull out to overtake and increase speed to continue up the next long steep hill, a car came belting down the hill towards me. I had no choice but to brake, pull in and crash into the back of the abandoned vehicle; the impact must have been at least 20 mph. I was belted up and suffered no injury at all except a stiff neck. I was even able to carry on to work after the shock had worn off. This followed a few yanks at crumpled metal on the front end with a large tyre lever; the front fenders had stayed virtually unmoved!

As a child in Stockport I was mad about cars, mainly I suppose, because we couldn't afford one for ourselves. One morning, my mum and dad said they had been in a fantastic new car belonging to their friends who had taken them out to a restaurant the previous night. 'It looked like one of those small Austins from the outside, but once you got in, it was massive!' said my mum, adding, 'It was called something like a Maxi.' It was unusual for Mum to get excited about a thing like cars.

'Rubbish,' I said as a know-it-all car freak. I knew every detail of every car, and had never heard of this Maxi thing. But not long afterwards my mum and dad's friend, Mr Want, turned up at our front door to take us for a spin in his Austin Maxi. It must have been a well kept secret because I'm sure I knew everything about cars at that time.

The Boot Room

It is nice to think that Austin's planning brief was to build a car with 'plenty of boot' and that the phrase was tragically misinterpreted by the designers.

Mr Want took us up to Adlington near Macclesfield, in his Maxi. There must have been seven of us in the car, but he allowed me in the front and pointed excitedly to the speedo as it registered 70 mph down one leafy lane. I loved the sticky up gear stick and the mock walnut fascia. I fell in love with Mr Want's Maxi from that day on, and dreamed we would have one one day. We often went up into the Derbyshire hills with Mr and Mrs Want in their Maxi which I think was a sort of orange colour.

We never did own a Maxi, but once my dad had a Marina which was similar inasmuch as it had that solid British clumpishness about it which others mocked but I grew to appreciate. When I was at Sheffield University, my room mate, John Weaver, had a beige-coloured Maxi which had the registration prefix 'CAR'. 'That should be done under the Trades Description Act,' said Simon Lloyd who is now a high flying lawyer. How wrong he was! John's Maxi ferried me and my pals all over Sheffield, and up to the Lake District for one memorable holiday. We paid him 10p a mile, I seem to recall.

Having lived in Manhattan for three years, I am now decidedly anti-car although I think I'll have to buy one to cope with an expected child and inevitable trips to the supermarket. We are looking for a Citroen CV which may not sound much like a Maxi, but in fact has equal charm, and similar down-to-earth characteristics which I can't quite define.

I wish Britain had stuck to making British cars like the Maxi instead of trying to imitate Japanese makes which are good: but they're not British.

I bought a four-year-old Maxi at 48000 miles for £1250 and sold it at 88000 miles for £100. Stupidly, I left an unexpired tax disc in it, so I guess I only got about £70 on the deal.

My first experience of a Maxi was when I drove a friend's - it brought me back in fine style from South Wales, since I was not used to a car radio at that time. Nor was I used to being so close to the blunt end of such light and responsive steering in a big car. In Tardis terms, it was an overgrown mind. The biggest drawback, I thought, were the sticky plastic seats.

Before and since, I have always had Fords, but I remember my Maxi with affection. It was, the salesman explained proudly, 'Harvest Gold', but he didn't explain about the split seam in the wheel arch which admitted farmyards of smelly water into the footwell. Eventually, they repaired it, and the only other bodywork replaced was the sills. Mind you, the front spoiler constantly had to be repainted with bitumen, and by the time I sold it, there were holes in the metal in front of the windscreen wipers.

I never had any trouble with the gearbox as such. I could perform very fast changes both up and down, especially into and out of fifth, though first and reverse needed a bit of effort to engage. The glossy knob on the gear lever was a congenial place to rest one's hand. It wasn't a fast car, and if you whistled while driving, the road vibration produced an extraordinary warbling effect.

Load carrying was spectacular. I was a teacher at the time, and collecting sheets of hardboard for school plays was no problem. Four eight foot by four foot sheets were loaded through the tailgate, coming to rest on the dashboard. A pupil assistant and I would then squeeze under the load, push up with our heads, and proceed back to the school. Since our necks were not very strong, and the Maxi in any case narrows towards the top, visibility was restricted to about 2 inches above the steering wheel, giving an outward appearance of a sort of mobile Marie Celeste. The original automatic car perhaps.

I am ashamed to say that the car on one occasion accommodated thirteen passengers for a short but genuine journey. It was a wonderful lorry.

Personally, I believe the Maxi conception was fine. Unfortunately the Maxi was marred by an awful gear box; I found this to be so in all three models that I had but especially in the first which had the original cable linkage. Still, in my later models, in which the cable linkage had been replaced by a rod linkage, gear-changing was far from sweet. I often missed the smoothness of the gearbox in my Land Rover which I ran for fourteen years before changing to my first Maxi; even with double declutching on second gear, the Land Rover gearbox was a joy.

By modern standards the 1750 engine was perhaps underpowered, but I found it fine - the Maxi was quite a performance car compared with the Land Rover. I do believe, however, that the 1500 engine was too small and that the lower powered Maxi should never have been introduced or, alternatively, it should have been withdrawn fairly early; one did not see all that many 1500s on the road.

It was, however, the body which I think was the Maxi's finest characteristic. It had the following features which I have never had collectively in any subsequent car:

1. A well-shaped aerodynamic rear such that the absence of a rear window washer was no inconvenience; I remember occasions when I did long drives in rain and on wet and salted roads with no rear visibility

problems at all. The downdraught over the rear kept the window completely clear and must have been an aid to performance efficiency.

2. The absence of a rear sill enabling easy withdrawal of heavy baggage or objects like filing cabinets.

3. The ability to drop the back of the rear seat, and of the front seats too to make a perfectly acceptable pair of beds on trips to NW Scotland. It was necessary to dump most of one's gear outside to enable two to sleep in comfort but on occasions when I was on my own and wanted an overnight stop without the bother of putting up a tent all that was necessary was to shove all the gear over to the right hand side; one could then have a very good night's rest. Many times did I use this facility, one that is, as far as I am aware, missing from cars currently available.

My parents possessed five Maxis, beginning with a blue 'J' reg. and replacing the car every two years with a new one. It was a brilliant car; there were three of us children and our umpteen friends. I don't remember my mother particularly enjoying driving the car. Something about the clutch! However, from a child's point of view it was a 'cool' car! Having just had my first child myself, I am fully aware of the strict regulations about child restraints in cars, so some of my memories as a child in the back of a Maxi in the 70s are a bit hair-raising. My mother once did the three mile 'school run' with at least twelve small children in the back.

We were the only people who had a car that had a 'bed' in the back of the car.

My brother spent a considerable period of his babyhood in his carrycot in the boot of the Maxi with the parcel shelf removed. I can remember him whimpering once, whilst we were giving a lift to some girls who were in a play my father was producing. They couldn't believe that there was actually a real live baby in the boot.

Lilford Park in Northamptonshire had what we and our friends considered to be an excellent adventure playground. Entrance to the park was per head. Somehow though, given the capacity of the Maxi to carry large numbers of small children, we never actually paid for everyone, with at least three crouched down, huddled in the boot. My life of crime in an Austin Maxi. To be fair to the people on the gate, taking the entrance money, I think they must have been glad to take any money from us and let us be on our way; an Austin Maxi, stuffed to the gunnels with marauding children must have been an intimidating sight.

If Dr Who drove a car he'd need one that was roomy, versatile, strong and with a design so elegant that it would not look out of place no matter what particular time zone he was in. It would have to be a sort of Tardis on wheels, it would be an Austin Maxi.

I have been the proud owner of a Maxi for seven years, I bought it with no MOT from a young bloke in Doncaster for £45. I put two new tyres on it, patched up the exhaust, got it through the MOT and have done nothing more than regular maintenance on it since.

I'm a self-employed painter and decorator and use the Maxi as part of my business. It hasn't got any flash sign painted on the side, it just looks like what it is, a hard working business car, struggling to make ends meet. It is covered in all sorts of paint. There was a point when I'd spilt so much paint on it by accident that it was starting to look a mess, so I got a brush out and swished as much colour as I could lay my hands on, all over the bodywork. I am quite proud of the result.

There is no need for a nice sign you see. When I knock at somebody's door to do a job, and the customer habitually peeks out of some upstairs window to see who it might be, they see my Maxi with ladders on the roof rack and paint tins strewn across the back, and it could only be one man.

Just as Dr Who wouldn't be the same without his Tardis, I wouldn't be the same without my Maxi. Sometimes, when there is a full moon, I think I am Dr Who.

I think that the Maxi was a great car. People will say that the Renault 16 was the design leader with family hatchbacks, but the Maxi spec of 1968 of 5 door hatchback with overhead camshaft engine, front wheel drive, disc brakes, 5 speed gearbox and fully independent suspension, was probably ten years ahead of its time at least. It is only with the Mondeo that Ford have equalled that spec.

I could sing the praises of Maxis for a very long time, suffice to say that in all the ten years and 130,000 miles or so, we never had any significant problems. The performance was impressive for its class, especially for the HL model, which I believe was road tested at around 100 mph max. The fuel consumption was good, always exceeding 30 mpg, except when towing. The interior space is legendary. The squarish back and low floor with no rear sill made it an ideal load carrier.

Perhaps the most irritating aspect of the design was the exhaust system. The movement of the transverse engine was allowed for by the introduction of a short length of flexible pipe at the end of the engine downpipe. This flexible section failed with infuriating frequency on both cars.

I am very long-legged and I would have liked a little more distance from the front seat squab to the accelerator, but that is a criticism I can level against most cars. However, the absence of a transmission tunnel did allow my left leg room, which was some compensation.

The spare wheel, slung underneath, should have been incorporated in a well inside the car, at the very least some cover should have been provided.

The front brakes squeaked. The single cylinder front brake calliper often meant that one brake pad would tend to remain in very light contact with the disc causing a very annoying squeak at low speeds.

Style! I must say that the purely functional looks of the Maxi had a certain appeal to me, but for the rest of the family its appearance was a disaster. The whole image of the car was fuddy-duddy; to use a word from the sixties, it was square. It lacked street cred, as today's youngsters might say. I would guess that that was one of the factors that resulted in sales well below the potential for the car.

After many years of driving a wondrous twin-carb Riley with walnut fascia and superb leather seats we wondered what we should buy to follow this hard act. It was suggested we buy a two-year-old Maxi - soft upholstery, twin carbs and plenty of lift-off.

My husband treated me to it new, August 1977. It went like a bomb, and still seemed so new when we could have expected to buy again in 1980, so we decided to keep it. This delay turned out to have serious repercussions, in that we disintegrated. I had developed ME in January 1982 and my husband fell ill with a malignant brain tumour. So, for 13 months I forgot I had a car, until at the end of January 1983 when my doctor told me I would be a different woman and I could go back to driving.

By that time rust had got in and the engine seized up. However, with the help of our capable mechanic, I was able to take to the road again; but you must imagine that by now I was tied to the old car which still had a great heart. Sadly, I broke my hip last July, which, combined with my ME, has kept me off the road ever since, apart from the day I test-drove a J reg modern automatic car which I thought I really should have, as ME really does not cure itself and I must think into the future.

Then I saw an advert in the local press; Maxi for sale, W reg, immaculate condition. I bought it for £165 and altogether have spent £570 on it. All the controls are in the wrong place - if I want to turn left or right, I get my windows washed. There is a funny little switch for the standing column and altogether far too much for me to do with my left hand. Added to that it takes a lot of kicking, swearing and grinding of teeth to get it to start. With both my former Maxis, it was a case of a bit of choke, turn the ignition key, into car and vroom. I have had new plugs, engine tuned and full service and it goes like a dream.

The Maxi is a superb car - a real workhorse, when I think of all the furniture I have moved for my old mother.

I bought one of the first Maxis when I went to work in Hong Kong in 1970. I think it must have been the only Maxi in Hong Kong. I'm not sure it is suited for the crazy driving out there but it coped well with the extremes in the weather.

There was this typhoon which made a late change of direction and scored a direct hit on Hong Kong. Unfortunately that night we had chosen to go to the cinema parking the Maxi in a multi-story car park adjacent to the harbour.

We were therefore ignorant of the typhoon's approach, but on leaving the cinema found chaos outside with scaffolding and hoardings blown down, all public transport abandoned, roads blocked and traffic lights out of order. We found our way to the car park only to find a power cut had ensured every light in the building was out and the wind howling in through its slatted walls.

It was quite a terrifying situation, but we were saved by owning a Maxi. To find any other car in that car park in the pitch dark would have been impossible, but by feeling the cars near where we had parked we were able to identify our Maxi by its unique shape. It was the only one on the island so there was no danger of us driving away some else's car! Thank goodness we hadn't bought a Japanese or German car which were very common.

After spending a very uncomfortable winter driving a frog-eyed sprite, my girlfriend's parents were delighted to discover that I'd sold the 'sports' car and bought a sensible saloon. This quickly turned to horror upon discovering it was a Maxi - the car with fully reclining seats, what fun. I think they wanted to lock her up right then. The year was 1974 and I'd bought a 1500 cable change Maxi which was to begin a long and enjoyable relationship.

If someone knew you owned a Maxi then you would sooner or later be called upon to carry a large item, usually cookers or fridges. On one occasion I was helping a friend move house and after careful measurements his upright piano was placed into the car lying on its back. Such was the snug fit that we couldn't get our hands under it to lift it out. I ended up taking the piano to work the following day and it was finally removed amidst much cursing and broken fingernails the following evening.

As a venture scout I was taking part in a large event in Cheshire. One of the events was a kite flying competition, but there was no wind - a problem when kite flying. Maxi to the rescue. Whilst a trusted friend drove the Maxi around the large grassed areas I sat on the roof flying a steerable stunt kite several hundred feet behind. The clear winners!

Holiday

I like to go to France
And romance
To have the time of my life
Without my wife
Oh Je t'aime my little fluffy
Austin Maxi

I'm not sure whether it was the marketing of the Maxi that did it, or the fact that its popularity coincided with the seventies boom in family camping holidays, but there is certainly something embedded in the Maxi drivers' psyche that compels them to take it on a 'little trip'. In the Maxi's heyday, Cross Channel ferries were sinking under their weight as they rushed like lemmings over the white cliffs of Dover in pursuit of the elusive cheap and cheerful foreign holiday. Country parks were littered with them, as urban Maxi drivers took to the hills in search of peace, tranquillity and the somewhat less than original photograph of their loyal open-booted family motor and flared-trousered son-in-law in front of a postcard natural beauty spot. Many thousands of miles of British bank holiday traffic jams have been led by some Maxi-pulled caravan crawling steadily to a site on top of a hill so steep that no one in their right mind would attempt to climb it; one that will remain for ever the kingdom of Maxi campers. I defy anyone who owns a Maxi not to be seduced by the unnatural urge to take it away.

I have been seduced by it once and it dominated the whole holiday. I wanted to make Valentine's day special for my girlfriend Isabel. We hadn't been away together as such before, so I booked a holiday cottage in Sawrey Knots in the Lake District for the duration of Valentine's week. With the winter winds howling, we loaded up the Maxi, jumped in possessed by Maxi holiday fever and set off with our usual disregard for practical measures and the normal day to day concerns of rational people. We were steaming up the M6 in a condemned motor car, flirting with disaster, overcome by the romantic notion of 'have Maxi will travel.'

My Maxi was in no fit state to go anywhere, the clutch had been slipping for about 3000 miles, there was a hole in the exhaust the size of a doughnut and it was carrying some mystery ailment which caused it to choke and stop suddenly after any prolonged period of throttle. I wasn't over-concerned, I secretly thought that the holiday fever was contagious and that the Maxi would somehow share the excitement of our first holiday together. Just the three of us.

It choked and stopped after sixty miles of M6 slog. It was inconvenient as we were keen to get to our destination but I was safe in the knowledge that once it had calmed down it would start up and be off again. We were grateful to be given opportunity for conversation. The hole in the exhaust and the vibrations of the car in general were deafening, it was common on long journeys that we would end up shouting at each other to make ourselves heard. We often climbed out of the car with a throbbing headache and a sore throat.

We slipped back into our own little worlds as the Maxi started up again and roared a bit further up the M6. The choking fit recurred with increasing frequency, we might get twenty miles before we had to stop, then ten, then five until finally we would be crawling along at thirty miles per hour in the slow lane until we reached our exit. A combination of determination, holiday fever and the little drawing of our quaint cottage in the brochure kept us going until we eventually hit the slow and winding Lake District roads. At this point we knew the mysterious choking wouldn't bother us again until our return journey, but the steep hills in front of us left us in no doubt that it was now the slipping clutch's turn to let us down.

Neither Isabel nor myself are great fans of walking and getting up at the crack of dawn with a map, compass and pair of wellies, then disappearing into a bleak and freezing wilderness in search of adventure. We got our adventure by attempting to drive up big hills before the clutch exploded. I got quite a buzz from driving around up there, not knowing what was around the corner and whether I'd even make it round the corner. After some of the more hair-raising hill climbs in the snow, Isabel was starting to worry a little and had visions of the embarrassment of being airlifted from our Maxi on a helicopter winch. I could see her look anxiously out of the window to the valleys below in search of a garage.

Isabel's concern, the faltering Maxi, the presence of snow and the absence of any hiking boots confined us quite happily indoors for most of the week. A quick peek through the curtains in the morning to see if the snow was still there, then we'd quickly close them and eagerly return to our attempts to rescue Minnie Mouse from the clutches of an evil witch on a computer video game. The Maxi sat patiently in the driveway, relieved that my initial enthusiasm for blasting round country lanes had worn off.

The slipping clutch, the hole in the exhaust and mysterious choking, had become predictable and lost their edge of danger. Their fearsome qualities were soon awoken however when I threw in an extra element; no petrol.

Our cottage was less than a mile from the Lake Windermere car ferry and Windermere village less than a mile's drive on the other side of the lake. During our stay there, the Windermere Theatre was hosting John Godber's *Up and Under II*, a play about Rugby League. Being a big fan of both the game and Godber, I was keen to see it, so we managed to lift ourselves from the warmth and comfort of our retreat and brave the winter weather for the short journey to the theatre. When we got into the car I did notice that the petrol was very low, but seeing as it was about a three mile round journey there was no need to worry Isabel about it. We rode across on the ferry, pulled up at the theatre and had a very pleasant relaxing evening.

When it was time to leave, we drove back to the ferry only to find that it was closed. Ah! To get back to our place we had to drive all the way around the lake, about thirty miles in total. I asked a passer-by the whereabouts of the nearest place I could get petrol at that hour and he casually informed me that it was in Kendal. And exactly how far is Kendal?

'About twenty-five miles.'

I shared the dilemma with Isabel. We decided to leg it to Kendal. The following half hour was the most hair-raising I have spent in a Maxi. We were flying down pitch black country lanes only wide enough for one car in the general direction of Kendal, with very few road signs and even less sign of any form of civilisation, with a winter's frost, an unknown quantity of petrol, a slipping clutch, a titanic roar, a mystery choking problem, a wing and a prayer. If we ran out of petrol here we would most certainly perish. The tension in the car was immense, Isabel was even prompted to say, 'Shouldn't we get a proper car, babe?'

The answer was obviously no, because in times of extreme pressure, in life and death situations, the Maxi invariably pulls through and so it was that we reached Kendal, filled up with petrol and two hours after we set off from the theatre to cover the short one and a half mile journey home, we made it back to the cottage and resumed our quest to free Minnie Mouse.

I've had enough frightening experiences on long distance runs within the confines of my own country to put me off taking the Maxi to a foreign destination. I am possessed with a romantic notion to take the Maxi overseas, but I know in my heart that it would never come back.

I am full of admiration for people who take the Maxi anywhere without concern, from the frozen tundra to the baking Mediterranean sun. Apart from the inevitable breakdowns on these trips, there is the added complication of the Maxi's simulation of a thermos flask. It is true to say that a Maxi is very cold in winter and melts at the first hint of any sun. With our island climate, the Maxi can cope, but in the extremes of the continent it's a different matter.

A couple of summers ago I went fishing with my dad. It was a scorching day, and in need of a bit of colour I took off my shirt and just lay in the sun for the whole day and thought that would do the trick. I was a lobster, burning red and so sore I couldn't move.

The next day, I was working at a car auctions office just outside Leeds, so I had to dress up in a suit. Putting on a tie just seemed to lock the burning sunburn in and I spent the whole day in agony. The sun was beating down, it was eighty degrees. At the end of the shift I ripped off my tie and moved cautiously to my Maxi. It was burning hot. The combination of the sweat pouring off me and the fact that I couldn't wind down the driver's window convinced me I was being given a taste of hell. I lit up a cigarette and set off home.

After about five or ten minutes I smelt burning and looked around the car anxiously and saw smoke rising from the gear stick, I was burning hot and my car had caught fire in the heat. I screeched to a halt, jumped out of the car and dived over some nearby hedges, certain the whole thing was going to blow. I lay there with my hands over my head for a couple of minutes before I dared look up. I peeked at the car to see the inside filling up with smoke. The penny dropped. I ran over to it, flung the door open and the carpet was on fire. My cigarette butt had fallen out of the ashtray and burnt a hole about the size of a dustbin lid into the carpet. I put it out and got in the car and drove off. I got less than a hundred yards when it ran out of petrol. Perhaps it was never going to blow after all.

The Maxi never really caught on with the native sophisticated Europeans and the expected rush on exports never materialised. Maxis did make it to the continent in their thousands however, and the following section contains stories of the inevitable problems encountered when this one-ton chunk of very British metal chugged around our federal partners highways like a mobile Union Jack.

The Trailer

Ever been stuck behind one of these? The caravan-pulling Maxi has played a large part in forming the stereotypical, pipe and slipper Maxi driver in the eyes of the British public.

This caravan is obviously sensitive to what kind of car pulls it, the manufacturers painting their preferred model on the front of it.

I Wanna Tell You A Story
Tour-A-Long-A-Max

When refuelling on holiday in France our S reg Maxi was always commented upon by young petrol pump attendants: 'Ah - le Maxi... '.

We had been warned about the notorious Périphérique around Paris, and received gratuitous advice about the unsuitability of taking such a plodding vehicle onto a road thought to be the preserve of the aggressively chic drivers of the Parisian rush hour.

Ignoring all discouragement, the Maxi entered the Périphérique unobtrusively by slip road. At first its performance was no more than totally competent - as one would expect. The precision of the responsive steering allowed us to navigate with perfect lane discipline past the first two or three slip roads, allowing impatient commuters to overtake and cut in at will.

It gradually became obvious that the softer-sprung, predominantly French-manufactured vehicles surrounding us were tending to rock and lurch awkwardly as they sought to change lane. No amount of horn punching could compensate for the sluggish road holding characteristics, or disguise the inferior acceleration of accompanying traffic.

The Maxi seemed to protest almost audibly at the restrained driving regime at first imposed upon it, by which I mean that we were proceeding at ten kilometres per hour slower than the speed limit. So I increased acceleration to just under the speed limit. Our front wheel drive now gripped the smooth tarmac of the French carriageway and we accelerated noiselessly into the fast lane.

As we changed lane, for safety purposes, or took the slip road towards our destination in Paris, the accuracy and firmness of the hydroelastic suspension system supported our passengers far more comfortably than the rolling and rocking attendant cars which we had outclassed in every single respect.

Oh - and we hadn't neglected the conservation of the earth's natural resources either. On this particular 'urban cycle' challenge, fuel consumption was close to 40 mpg, typical of the Maxi.

We had a G registration Maxi in the 1970s, and to us it was a very capacious car. In fact, we used it to transport all our personal belongings to Germany when we went to work there as teachers in 1973. Imagine our annoyance when a German friend misread 'Maxi' as 'Mini' and persisted in calling it that.

However, it really came into its own when we started using it as a travelling hotel. One Easter we decided to drive from Central Germany down to Southern Italy, to visit Pompeii, and then to drive back - all in a week. To save precious time and money, we slept in the car, opening out the front seats and the rear seat to form a double bed. When we reached the Swiss Alps, the weather was freezing cold, frosty and snowy, so we parked down a little path, fixed the 'bed' and got into the sleeping bag, flinging our underwear onto the parcel shelf. The only problem was that, when we woke up the next morning, we saw our knickers frozen to the rear windscreen and covered with frost.

All in all, the Maxi was a car we still sorely miss.

After travelling all around France in our Blue 1500 Maxi on camping holiday, my wife and I were going north from Grasse along the Route Napoleon with torturous steep slopes in sweltering heat. The car which was heavily laden with tent, duty free, French mustard, bushes of garlic, the whole French job, began to falter near the summit where there is a lay-by to admire the impressive scenery down to Chasse and beyond to the coast. I was convinced that the car, like us, wanted to stay. The problem was minor. In fact I can't even remember what it turned out to be. The following year, in a different red 1750 Maxi, we had been on another circular tour of France but were returning north along the same route through Grasse when the car began to slow and struggle, forcing me to pull in at exactly the same lay-by where I had been stuck the previous year. This time I remember the fault distinctly - a slack distributor, despite a recent service. It was twistng back and forward so affecting the timing. At the time I was unable to diagnose the problem and was convinced that fate was lending a hand at suggesting I should be rich enough to live on the Côte d'Azur. After a brief stop-off, we went on again and though the car's performance was sluggish we still got home.

I am a self-employed motor engineer and was first introduced to the car when a customer brought it to me for repair. He had bought it from George Clarke, an agency for Leyland cars. This particular Maxi was registered under George Clarke's name for his personal use. I had admired the condition of the car and had mentioned to the customer that I was interested in the car should he wish to part with it. The car was meant for us because at the same time as the car came on the market I was arranging a holiday for five of us in France and the Ford Escort that we were running at the time was a little cramped for three in the back seat.

The Maxi easily accommodated my wife and me, our two teenage children plus my daughter's boyfriend with the combined luggage. We set off for Portsmouth in the pouring rain for the overnight crossing and arrived at Cherbourg in France before dawn. The diffusers were fitted to the headlights by the light of the street lamps and we sneakily manoeuvred it so that the boyfriend was the first to drive on the right-hand side of the road even though the poor lad had passed his test not very long ago. Well, these young people take to new things much better than us middle-aged folks!

The driver's reactions were put to the test on the way south to the Vendée when travelling through what appeared to be a deserted village. Suddenly, a car reversed out from a garage on our side of the road straight in front of our path. We escaped by a hair's breadth and, apart from finding the roundabouts a bit strange and my wife going through a red light, the rest of the driving was trouble free.

The Maxi has conveyed all the worldly goods of my children to and from colleges, boxes to craft fairs, oil drums to garages, drum kits to concerts and children to the seaside. My wife and I have slept overnight in it as has my daughter. The latter time was on the way home from the holiday in France for which we bought the car and my daughter was very nervous about going off to sleep in case someone looked in at the window at her. My son and I slept in a tent on the grass of the lay-by and my daughter's boyfriend was in another tent. My daughter awoke to find a face peering in at her as she had feared - she screamed before she realised it was the boyfriend!

It was our very first visit to France. I knew that our frame tent was not top of the range but the weather in the south of France was renowned for its sun and clear skies. We pulled in to a small campsite at Courthezon, near Orange, parked our Maxi, pitched the tent and left to find somewhere to eat. It was only a short walk to a nearby restaurant where we ordered our meal. About half way through there was an enormous storm, followed by an ear splitting drumming noise. A lot of children who were eating with their parents ran outside and brought in giant

hailstones, my memory swears that they were almost golf ball in size. I leaned over to catch the arm of the waitress and hotel Madame, asking her in abysmal French if there was any possibility of staying in the hotel overnight. The reply unfortunately was negative.

By the end of the meal I was convinced that the tent would be thoroughly waterlogged. The hail had given way to torrential rain as we left the restaurant. We arrived back at the tent to find that it had only just started to leak in but the ground was swimming in water. We put everything we could on camping chairs covered in sheets of plastic, and emptied the Maxi of everything so that we could sleep in it. I was very glad of the second bottle of wine that night. My wife did tell me that it was rather unpleasant. I do remember the rain and hail drumming on the roof, but slept soundly for a good few hours.

I was awoken in the early hours to the sound of a siren going off and the nearby rush of running water. What made matters worse was that just before we left England we had watched the latest in the then rash of disaster movies. This one had involved a dam bursting which had been heralded by the sound of sirens! The thought of the car being washed away was the least of my problems, by then the red wine was giving me other things for my head to be concerned with. The next morning was fine, we bundled the tent into a large plastic sack and headed for Cannes, about 100 miles away. When we tipped out the sodden tent the owner of the camp site, when he found out where we had stayed, muttered 'Ah the Mistral,' and promptly told us that they had had no rain for three months.

Whilst on holiday in the Lake District many years ago, my wife and I were enjoying a quiet drink in the pub at night, before departing the next morning for a long drive back to Essex. Leaving the pub, I switched on the ignition to return to our 'digs' There was only a click and the lights dimmed. 'Here we go again,' I said and checked by torchlight. All connections were ok and lights were on, bright and full. I turned the ignition, click, and all was dimmed.

I phoned the AA and they could not get to me from Lancaster for three hours. Meanwhile, my wife had gone back to the bar for a drink as it looked like being a long night. One of the locals who had heard of our plight said, pour boiling water over the positive lead of the battery terminal - he had a Maxi himself. Well, it worked and the engine fired into life, the lights came on and we were away. The problem never recurred again.

We bought the Maxi new in 1980, and still own it. My daughter now drives it and enjoys being 'different' in a non-standard car. The car has been maintained throughout its life by Blackmore's Garage, a small garage in Cardiff ran by brothers John and Bill Blackmore who were always complimentary about the way it has maintained its condition. It's not as though we only drive it to the shops on the odd occasion, I am immensely proud of the way it took us on trouble free trips to the Faroe Islands, the Outer Hebrides and Iceland in the 1980s.

The Faroe Islands are particularly rugged and mountainous. The islanders have developed a road network by burrowing through the mountains, since in many places they would be unpassable in any other way. My wife who is claustrophobic, was none too pleased at having to cope with the tunnels. We stopped before the entrance to one tunnel for her to pluck up the courage to go through. The map showed that it was about 1½ miles long but straight. When we entered, we found that it was a single lane tunnel with occasional passing places, but no lighting and no lining to the walls. To my wife's relief we could see the other end as soon as we entered. Imagine her horror, when, half way through, the light at the end of the tunnel disappeared. I shall always remember her grabbing my arm in fear that the tunnel had collapsed. However the end of the tunnel soon came into sight again - the road surface had dipped to such an extent at that point that the Maxi had gone completely below the sight line of the tunnel exit. We had to return by the same tunnel and again Mary had to draw on her resources of courage to contemplate going through the tunnel. This time we were followed in by a large lorry. When a car came the other way it was his priority so we had to pull the Maxi into one of the passing places. The lorry followed us in but the space was very small and we had to spend quite some time manoeuvring up against the wall of the tunnel to squeeze the lorry in for the car to pass. By the time we got through Mary was a nervous wreck, but I shall always admire the courageous way she had suppressed her fears for the rest of the family to enjoy a day in the Maxi on Vidoy Island - truly one of the most spectacular locations in Europe. We kept the green temporary licence sticker - Fyribil Loyvi - in the windscreen afterwards as a reminder of a wonderful trip. Some of it remains 13 years later.

We could recount many stories of our trip to Iceland. All I will say, though, is that the Maxi drew many a curious stare. Very few of the roads in eastern Iceland are tarmacked, yet the Maxi coped with dirt tracks and grit roads admirably. We were very heavily loaded with camping equipment for a family of four for a month. I had taken the precaution of getting the suspension raised for maximum load, yet it was still very low down. Many of the roads in Iceland are more the preserve of four wheel drive vehicles. At one point the track we were travelling on was so rocky that Mary had to walk in front to move some of the larger stones out of the way. But we were all proud of the way the Maxi came through without a complaint or a fault.

We looked at a new white Maxi with blue upholstery and fully reclining seats in October 1971. This was definitely going to be our car, and so it was for ten years, formative years for our three children with our old Herald Estate and our new Maxi serving us for a long time.
 They were mostly trouble-free years, but not entirely so. In 1976 I had a two-week commitment in Grenoble. That was a glorious August in England but in Grenoble it rained nearly every day and moss grew in the window crevices of our Maxi. That was not the big problem, though, as early in 1973 I had suffered a broken constant-velocity joint. Would the other one fail, I thought, as we crossed Rosslare to the Havre in 1976. As we approached our first night stop in Chenançon in the Loire Valley, it would and it did! Accompanied by nasty expensive sounding noises I drove gingerly in a sweat through to Grenoble on the third day. Then I found the BL agent in Grenoble and, using an AA voucher, a repair was provided soon. 'Au revoir,' said the garage man, but I insisted on 'adieu'!
 There was no more trouble of a mechanical nature to worry us, and the new joints lasted the remaining life of the car, but we were due to go to Switzerland, where my elder daughter was engaged as an au pair for a month or two. That was a day's journey, pretty straight on the map, except for a wiggly bit near the end. This wiggly bit turned out to be the Furka Pass, and for the first and only time of my life I experienced snow in August. However, the road was open and well engineered and caused us no trouble as we joined hardy-looking Swiss militia on their weekend exercises. Even so, loaded down, I took it easy and did not accept the challenge of the lightly loaded 2CVs as they joined us on the straight bits.

Our first Maxi was our first new car - 'Tundra green, UNK 989M'.
 It was comfortable, powerful, and never let us down. We camped in it in Scotland, complete with 'anti-midge curtains'. It was loaded to the gunnels during winter logging expeditions and actually managed to hold an entire three piece suite on one occasion. On the night Mr Heath's boat

foundered off St Bees Head we were there and went on to the shore-side camp-site, and attached a tent to the rear of the car. It was force 10. We slept OK but rain forced through the new tent and infiltrated locked suitcases, and people. Father was dry - in the Maxi. At daylight we saw the people next door shoving what was left of their tent into the bin and driving off! We had a hot breakfast, and proceeded to dismantle our large tent from inwards - shoving it into the tailgate. What a lovely car! We then drove it over bumpy sand nearly to the water's edge - you never saw such a sea. No wonder his boat sank!

Our next was 'SRO 22R'. Sandglow, purchased from my brother, and nearly new. I didn't like the colour, but it was a good car. She took us across France to Alsace Lorraine where we played snow balls in June on the summit of Ballon.

We also toured a caravan at home and abroad, and found the front wheel drive useful for pulling other vans out of mud. Our next venture was Norway. The roads were great, no speeding, as the Norwegians have a nasty habit of cruising main roads in helicopters. Instant fines are levied and I believe they take you off the road too. We went up in the high mountains,no huffing or puffing, and then above through roads with cliffs of snow in June on either side of the car, three times its height.

The ferries across the fjords were lovely and the cars sat on the lower deck, whilst people were above on a balcony. As our bonnet was still hot, I spread out all my washing on it, including personal items; not realising I had a grinning audience of Germans and Norwegians above me! Maxi owners never blush.

We replaced 'SRO', rather sooner than we intended as once more my brother was selling his. We bought it, MBM 284V, and ran it for years. It now lives in North Watford with my daughter, making periodic visits to our house for outpatient treatment by dad and brother. Her paint is in good nick and she has just passed the MOT. Apart from a few knocks, and groans which have all been fixable, she is running well, and has just gone off to Cheshire for a few days' holiday.

On the sad day when Maxis were no longer made we had to resort to a Ford Sierra Estate. I could have cried, it's not a patch on a Maxi and it's so nice to see good old MBM drive in to the front garden for a check up.

To me its sleeping facility was one of its great virtues. At the time I owned them I was doing a lot of camping and hill walking in Wales, Scotland

Lake Windermere's Maxi Fan

This is my Maxi, and if you look very closely, behind the steering wheel is my girlfriend Isabel.

I had fallen into the trap. Not satisfied with the beautiful view, before I took the photo, I had to get the Maxi and Isabel in shot so I could get that little extra. I have many such photos.

I believe Polaroids were invented so we could keep such shots our little secret.

and the Lake District. Though I had a tent my wife and I often used the tent for storage and slept in the Maxi or used it to get a few hours' sleep when travelling great distances. On one occasion I was camping alone and walking in the Lakes, and planned to camp in Patterdale in order to climb Helvellyn the next day. Unfortunately a coach accident on Chestnut Hill overlooking Keswick caused a long delay and it was pitch black and very late when I got to Patterdale which is between Keswick and Grasmere. I therefore decided to forget the tent and sleep in the Maxi. I woke up some hours later to find the moon had risen and three 'judges' with black faces and 'wigs' fringed with moonlight were looking at me as I sat up. Thinking it was judgement day at last, or that the couple of pints of Border Ale was stronger than suspected, I tried to weigh up the chances of a dash for the adjacent road when one of the 'judges' gave, not a verdict, but a bleat!

It was 2 July, 1986, and we were on the Autoroute from the Italian Border at Ventimiglia heading for a campsite at Aix-en-Provence. We had been away from home for five weeks, trouble free, and it was a beautiful day cruising along at 60mph. We were on the outskirts of Cannes and in the distance I could see the signpost for the turn-off to Grasse about halfway up a slight incline. I suppose I could have made it in fourth but, ever considerate for my Maxi, down with the foot, gearstick up to third - and nothing happened. There was no crunching of gears so it wasn't the clutch. The Maxi just kept going on in fourth and slowed all the time as I suppose in my panic I had eased my foot on the pedal. With the van on the back I had to give in and stalled to a halt pulling over to the hardshoulder. Out with the warning triangle. Hasty word with the wife - No I didn't know what was wrong. Dash to phone and explain position in my kind of French - 'Voiture est mort' - 'Maxi Bleu' - back to wife for a moment reflection, at least the sun was shining I could get out a chair and have a G and T and a smoke. No such luck, I had to run back to the phone - I'd forgotten to say that I had a caravan derrière! Twenty minutes later the breakdown truck arrives accompanied by the motorway police. Their English was less than my French but it didn't take long for them to realise it wasn't a roadside repair. We then suffered the indignity of sitting up front with the driver, Maxi behind us and the caravan trailing along behind. We took the turn-off for Grasse and almost immediately pulled up at the small two-ramp garage at Mougins. Drop the van off round the back and then roll out the Maxi down the ramp and on to the forecourt. The garage mechanic, little

English, indicated that I had to pay him for the tow and the repair - by cheque with Eurocard - and returned in four hours. It was ready and waiting - the total cost - Tow 602F.F. Repair 100F.F.

You may have guessed that the sheer-pin for the selector rod had in fact sheered! I have to give praise to the mechanic because not only did he replace the pin he also threaded a wire through it. I don't know the French for belt and braces.

We were caravanning in Spain on the main road high above the town of Blanes. The campsite we had looked at was too filthy for us to stay and the alternative was through the town and a mile or so on the other side. Leaving the camp I could see a road to town if I crossed the local rubbish dump. There was a path of sorts, my wife objected but the Maxi didn't and that's what mattered. Across the dump I could have taken a left turn and then a whacking great detour to town but I decided instead to take a short-cut down a very, very steep hill. No sweat. The Maxi had good brakes and in first gear would hold back the van no bother. Down, down, down, we went until the road narrowed ending in a small square with only an alleyway leading to the Main Street, Blanes. There wasn't even room to turn the Maxi. So, unhitch the van, push it back with the help of locals, back Maxi and turn, hitch-up van - then what? To my horror I found that the hill down was one-way and I was obliged to return by a parallel road which looked even steeper - I guessed 1:6 - and there was that small road crossing it now near the bottom which may or may not have traffic, well here goes. Into first, down to second - 20mph - be careful at the crossing then foot down. I decided to crack it in second. 100 yards from the top I had to find first gear - not an easy task at anytime - got away, down to 10mph then the road steepened and was covered in loose chippings.

The brave Maxi up at the front with the weight on the back just couldn't get a grip on the stones and, not for the first time, stalled to a halt. Leave in first gear, handbrake hard on and a quick dash to the van for chocks for the van and car. I was horrified when I looked down the hill and mortified when I looked the 60 yards or so to the summit. A kindly Spaniard called meccanico, pointed to a bar and indicated telephone. Sure enough, in under the hour, police arrived with a Land Rover and winch and, with the Maxi restarted, we conquered Everest. I feared a hefty fine and even wondered if the Bail Bond might be required but my wife smiled, the police saluted and suggested the mechanic would accept our monetary appreciation. Next day we returned with the Maxi having left the van on site taking a bottle of brandy to our Spanish friend and having another look at the road which had now steepened to 1:4. wondering all the while if we were right to assume that the Maxi could take us anywhere.

T he highlight of my Maxi ownership came with 'Mad Max', RPM 703X, a dark blue one which had kindly been lent to me by my parents, as I ventured abroad for my year out from University in France. He definitely made his mark on the continent and became a constant talking point. I lived in Annecy with two other British students and Max was soon accepted as part of the household. The flat that we had rented was unfurnished. Max immediately proved his worth by carting borrowed beds, chairs and tables from one end of Annecy to the other. Many of our French friends were known to exclaim, 'But you can't possibly fit all that lot into there!' Yet they were underestimating the unlimited amount of space in the back of a Maxi. In fact, this was the one point which used to redeem the Maxi in the eyes of my school-friends. They saw that the Maxi certainly held great potential as a 'pashwag'!

I worked at a comprehensive school in Annecy, teaching English to French pupils. Whilst the French teachers had all become adept at parking their 2CVs in the very small courtyard in front of the school, I encountered a few problems with my 'tank' of a Maxi! Being totally inept at parking, I found it very distressing to have to carry out my parking manoeuvres in front of 1800 jeering pupils who had been attracted by the inevitable Maxi clanks and rattles. The Maxi did however increase my standing in the eyes of my pupils who rightly saw it as a car full of character. They would queue up after school hoping for a lift home in the strange British contraption that 'Madame' drove. This was in direct contrast to my friends at home who would always offer to drive rather than suffer the indignity of arriving somewhere in a Maxi! My phone-calls home from France invariably contained the line, 'The car's making another strange noise again.' My father would then spend a while trying to establish whether this was one of the normal Maxi noises or whether it was a new variety. During my year abroad I became an expert on Maxi sound effects - the strange, the very strange and the totally weird!

Max was by no means a spring chicken when he set off for France and I was not surprised to encounter the odd problem with him. Fairly soon after arriving in Annecy, one of the teachers at my school said that I must drive up Semnoz, one of the nearby mountains, to see the magnificent view at sunset. Having set off late and having then lost our way, my flatmates, Chris and Rachel and myself, were resigned to seeing Semnoz by moonlight rather than by sunset. Little did we know at this stage that we were not going to see the top of Semnoz by any light at all. Max had puffed and panted valiantly around at least fifteen hairpin bends but then suddenly failed me on a steep uphill stretch. He was completely powerless and would not respond to either persuasion or anger. Having learnt from my mother's irresponsibility, I am a firm believer in the fact that by administering oil and water, you can cure any car problems. However, upon checking the oil and water, I found that neither was low. We were stuck in the dark, surrounded by forests and with no other cars or habitation in sight. Since oil and water were the extent of my mechanical know-how at that stage, I was well and truly stumped. We succeeded somehow in turning the Maxi round in the road and pointing him downhill. We then clambered back in and chanced our luck. The engine was still going but I just couldn't get any acceleration. This should have given me an immediate clue as to the source of the problem but I was too concerned with getting us down the mountain to worry about cause and effect. Max rolled obligingly down the mountain for us and was soon approaching the bottom. He cruised happily towards a large junction and we all sat with bated breath praying for the traffic lights to be green. Luck was on our side and we had just enough momentum to carry us safely through to the other side of the junction where we came to a timely stop right outside the police station.

 The policemen on duty were rather amused by the plight of 'les anglais' and were quite cheerful in telling us that we would find no Maxi garage open at that time of night - or at any time for that matter! They suggested that we should push the car into the police car-park and pray for a miracle to happen on the following day. You would have thought that by this stage we would have been glad to have Max off our hands. For some inexplicable reason, we were reluctant to desert Max and decided to push him all the way back to the flat. Having only been in Annecy for a short while, we were under the impression that our flat was situated just around the corner - little did we know that it would have been at least a two-mile corner! Luckily, the policemen spotted us pushing Max in the opposite direction to that which they'd indicated for the car-park. After much waving of arms and protestations, but no visible signs of offering to help us push, the police finally persuaded us that the car-park was the best option. We were most grateful to them when we discovered that the two-mile walk back to our flat was enough of a slog without the extra exertion of pushing a Maxi along too.

The next day, when I had a break between lessons, I rushed along to the police-station to find Max waiting forlornly in the car-park. I started him up and hoped that the prayed-for miracle would come into effect. No such luck! Max was still limp and lifeless. Rachel and I had soon learnt the advantages of trying on the '3 Fs' whilst in France: Feeble, Foreign and Female! The addition of a Maxi to this formula was a dead cert for success. In true style, I opened the bonnet of the Maxi, got out my Maxi manual, pretended to pore over a particular page, which happened to be about tyre pressure, and looked as '3 F-ish' as possible. Sure enough, a friendly French bobby appeared at my side asking if 'Madame' would like any assistance. The engine of a British Maxi was obviously as alien to him as it was to me but he soon became intrigued by it. He became so absorbed by it that he almost forgot about the problem in hand. However, when I'd pointed out the symptoms to him, he was quick to diagnose a slack accelerator cable. He fiddled around somewhere inside the engine and declared that he had carried out a temporary measure to bring Max back to life - i.e. done one of those familiar botch-jobs at which my father has become so proficient during the years! He then told me that I should get Max to a proper garage as soon as possible. Seeing him as the saviour who had rescued Max from a potential terminal illness, I almost flung my arms around him as I thanked him with great gusto. I believe to this day that I have dispelled the myth for at least one French person that the British are cold and reserved.

Since nobody had been able to give me any indication of a nearby Britsh Leyland garage - not surprising really - I simply drove in the direction of our flat and into the first garage I found. The fact that it happened to be a very smart Peugeot garage did not deter me in the least. I drove up to the large, glossy automatically-operated workshop door which opened slowly and impressively to let me into the Aladdin's cave beyond. Gleaming Peugeots of all shapes and sizes were lined up on either side of the workshop. The arrival of a British car, and more particularly a Maxi, was enough to draw the attention of all the workers. I had to resort to the ever-faithful '3-F' technique again and approached one of the mechanics, saying, 'I know that this car is not quite a Peugeot but I was wondering if you might help me...' I was immediately surrounded by blue-overalled workers who were dying to see this 'odd' British contraption. After exclaiming for ages about the originality of Max's engine, they finally tightened the accelerator cable for me and refused to accept any payment. Either the novelty-value of the car or, the '3-F' treatment had definitely done the trick! I returned later on that day

Maxi Photo Opportunity

David Collis Smith certainly knew how to take good Maxi holiday photos. The top one is a particularly good shot of the Maxi in front of the spectacular Allalin Glacier in Switzerland. David is so cool, in the bottom picture, taken in the Brecon Beacons, he can chat to his wife Brenda, tie his shoe laces and still manage to get a great shot.

with the statutory bottle of whisky which we'd had the sense to bring over from England for just such occasions - it seemed to amply compensate for the lack of payment! I became quite a regular at the Peugeot garage during my time in France and was always greeted with great delight, 'Oh la la, it's the one with the strange machine again!' My alternator-belt was tightened in exchange for a box of shortbread, and so it went on....

My next test came when I was driving back from a week's skiing holiday with my family who'd come out to visit me. During our last night at the ski resort, there was a massive snow-fall. The roads were treacherous but Max remained unperturbed. Despite not having winter tyres, he held the road really well as he followed the snow-plough down the mountain. Disaster struck at the bottom when the snow had turned to pouring rain and my windscreen-wipers packed up. My sister had to act as a manual windscreen-wiper on the hour's drive back to Annecy in the worst rain possible. My father was luckily on the spot to give his verdict. He said that I'd need a new cable, one of the few things that was not to be found under the seat. My grandma was coming to see me within a couple of weeks so my father agreed to send her out with the appropriate cable. In the meantime, we became obsessed by weather forecasts, knowing that it was only safe for Max to go out on fine days. Grandma duly arrived with the cable, together with a new wiper blade to replace one that was very worn. I will always carry this image of her arriving at Geneva Airport with a windscreen wiper blade proudly sticking out from the top of her basket!

This time, I decided to put the 'Maxi Manual' to its proper use and not just to use it for show. I spent hours trying out various tools and desperately trying to fit the cable. I am not mechanically-minded so felt a great sense of achievement when I got my windscreen-wipers back to life again. I hadn't been able to get the adjustments quite right so the blades only covered three-quarters of the screen. However, I viewed this as a mere technicality in the light of my great feat of mechanical engineering!

Grandmas and airports have a particular significance as far as the Maxi is concerned. My flat-mate, Rachel, asked me to pick up her Gran, Nanny Barber, from Geneva Airport one Friday lunchtime. We set off slightly late, having both been teaching early on in the morning. A few miles out of Annecy, Max did his favourite trick and suddenly died on us. We managed to push him into someone's drive-way and tried to discover where the problem lay - we had become quite expert by now at locating Maxi faults. Unfortunately, it did not appear to be any of his 'normal' problems so we were stumped. Time was running out and we couldn't leave Nanny Barber stranded at the airport. There was only one option.

We asked the owners of the house if we could leave Max adorning their driveway and resigned ourselves to hitching. We were lucky to get a lift fairly quickly with two young men who were going to the centre of Geneva. The airport is a fair way away from the centre. Once more, the '3-F' plan, together with the sob-story about Nanny Barber, did the trick. The young men kindly delivered us to the entrance of the airport and we were just in time to see Nanny Barber coming into the Arrivals lounge. We had planned to get a bus back to the centre of Geneva, and then a coach back to Annecy. Upon hearing of Max's plight and our unexpected hitch to the airport, Nanny Barber immediately insisted upon hitching back too. She saw it all as a great adventure and was game for anything. Rachel and I stood with our thumbs out at the side of the road. Nanny Barber sat on her suitcase and proudly held her thumb out too. It took us three different lifts to get back to Annecy but we made it, Nanny Barber and all! She was almost sorry to hear later on that Max had started straight away when I went to try to sort him out - I think she'd been secretly looking forward to a week's hitching.

As was borne out by this episode, the Maxi is often unpredictable in a positive kind of way. As well as dying on us for no apparent reason, he'd sometimes start up again just as inexplicably. My sister always remembers what should have been a simple trip to the hypermarket when she was out visiting me. Rachel, Chris, Jo and I piled into Max to go and do our weekly shop.

About half-way there, he had one of his fits and spluttered to a halt on the dual carriageway. A kind Frenchman pulled up and had a quick look at the engine. Unable to understand the intricacies of Max's engine, he offered to tow us somewhere. We asked to be towed to the hypermarket! We then deserted Max, who was by now not exactly flavour of the month, and went to do our food-shop as if nothing were amiss. Since we were throwing a large party that week, the food-shop was rather larger than usual, including several crates of beer. We loaded this all up into the Maxi, still trying to ignore reality. We then got in ourselves and waited with bated breath as I put the key into the ignition. Max roared to life as if there had never been anything wrong and safely transported us, and most importantly the beer, back to our flat. He had once again redeemed himself in our eyes.

I have always sworn that if I ever get married, Max will be my wedding car. I might turn up to the church very late and covered in oil but it would definitely be worth it. Max has been sitting in our car-port for the last year or so, patiently waiting for this great occasion. I have to confess that, having been a Maxi family for so long, we have become traitors to the cause for the last couple of years and are now a Maestro family. I like my Maestro but have never become quite so attached to him as I was to Max. There have been several suggestions of sending a now very aged Max, the last in a long line, to the scrap-yard but I have always strongly vetoed this idea. Now though, my family has decided to trade Max in, in part exchange for a Rover. It may be one step up from the breaker's yard but it still does not meet with my approval. It was a very sad morning for me the other day when, being the only person insured to drive Max, I was faced with the unpleasant task of taking him along for his MOT, I felt a great sense of nostalgia as I got inside Max again and was faced by his smooth wooden dashboard. The plastic dashboard of my Maestro is like a Fisher Price toy in comparison. I drove gingerly to the garage, warming to the once-so-familiar wheezes, splutters and clankings. I was sorely tempted to elope with Max there and then but my father had come with me to see that I saw my unwelcome duty through to its bitter end. Since I took Max along before work, the garage was not yet open so we left Max in front of the entrance. Apparently, my father received a phone-call later on that morning to say that Max would not start and that he was totally blocking the way in to the workshop. Max showed his perverse streak right to the very last! I now have to face the prospect of an empty car-port at my parents' house and the fact that I shall never get married in true Max style. A car is supposed to be something to get you from A to B. Max did not always fulfil even this criterion but at least we had fun and games on the way.

Performance

Oh Maxi, Maxi, don't break down
With your shiny hub-caps and your paint so brown
Oh Maxi, Maxi, I think you're great
Don't let me down, my old mate

Opinion on the performance of the Maxi falls into two camps. There is the bunch with twinkling eyes who think it is the epitome of British craftsmanship and will run forever; and then there is the other lot who view it as the brainchild of Satan. I fall somewhere in between. My Maxis have broken down many times but I tolerate the inconvenience of it all by the virtue of their incredible capacity to get over it without too much fuss.

Fresh out of college, my friend Karl and I decided to do the Dick Whittington trip and headed off to London in search of our fortune. We packed a bundle of naive optimism in our kitbags, threw them in the back of the Maxi and were off, steaming down the M1.

It was a jovial, 'look out London, here come a couple of Northerners to sort you out' type of trip. It was a Saturday, and being a big Sheffield Wednesday fan, Karl was keen to tune into the radio as his team were playing big rivals Sheffield United. In-car entertainment - in terms of audio equipment - was a concept developed much later than the Maxi. They all have radios, but if the vibrations didn't kill the sound, the engine surely would. We tuned in anyway, and if we kept the speed down below fifty we could just about hear the commentary over the drone of the engine. There was much jumping and swaying as Wednesday got an attack together, and cursing and swearing when the radio reception temporarily lost the battle with the hole in my exhaust.

Then came the most dramatic moment I have ever encountered in all my time behind the wheel of a Maxi. In one incredible second, Wednesday scored, Karl leapt into the air and the Maxi, without any prior warning, quite spectacularly died.

So there we were, on the hard shoulder, stranded. The naive optimism that had catapulted me to London in search of fortune with the unlikely combination of an Austin Maxi, a Sheffield Wednesday fan and thirty quid in my pocket had extended to an absence of any breakdown cover. I did know that if I was to be recovered from the hard shoulder it would prove very expensive resulting in me being sent back home without the swagger and big fat wads of cash that I was hoping for.

I got out of the car and paced up and down a bit in a futile attempt to find inspiration to get me out of this mess. Karl did his bit and joined me. His face showed all signs of concern and sympathy, but I could see his eyes were dancing about and secretly chanting, 'One-nil'. No words were spoken, no solution reached, it was a real life Laurel and Hardy film.

We climbed back in and as though it were the last act of a desperate man, I put the key in the ignition and turned. It started. We set off down the hard shoulder, built up a bit of speed and pulled out. It was as if nothing had happened. We made it to London, drove around for a few days, didn't find a job, except Karl who was miraculously successful in obtaining a job selling flick-knives. I managed to convince him to turn it down. We quickly and surely ran out of money and drove back, the Maxi showing no signs of concern. I never figured out what was wrong with it, my only action was to leave the radio off for the rest of the journey, for although I didn't know what was wrong with the car, I figured it would be a good idea to keep Karl still.

I hadn't had the Maxi very long at that point and my knowledge of carburettors and camshafts and all that stuff was zero, so me and any car that mended itself were going to get along just fine. It didn't end there, several times I have broken down, opened the bonnet, peeked inside, closed it again, turned the ignition and drove off.

On one occasion things almost got out of hand. I was driving round the M25 towards Guildford and I'd been hearing some rattling sound for a couple of hours. I was cautious but not over-concerned when suddenly my brakes failed. The pedal was just flapping around. I still had forty miles to go, but once again no breakdown cover, so a touch irresponsibly, I carried on. With careful use of gear changes and plenty of handbrake, I made it safe and sound.

The next morning I took it to a local garage and explained my brakes had failed, and by the way, there was a rattling sound. He took off my hub cap and three wheel nuts fell out. My wheel was hanging off.

'There's your rattling sound.'

He put it up on the ramp and proceeded to point at all sorts of things and without much hesitation told me the car wasn't roadworthy and should be scrapped. I explained in a simple way that scrapping wasn't an option and if he would be a good chap and bleed my brakes, I'd be off.

More than his jobsworth. He couldn't allow me to drive a condemned vehicle from his premises because if I were to have a spot of bother in terms of fender bending, then he would be in deep trouble. I threw in a disclaimer idea, simply saying, 'It was all my fault', I would sign it and that would be that. He bought it, did my brakes, and I was off. I ran the car for a further six months and the problem never repeated itself.

In all my time with the Maxi, it has only ever had one problem at any one time. It is true that as soon as I mend one part of the car, something else will go. This situation led me to the philosophy that I should never repair anything until it is absolutely necessary. The Maxi was strong enough to ride over many ills, and sometimes minor irritations like wheel bearings and exhausts falling off could be left untouched for three or four months. Once a small hole appeared in the exhaust resulting in a tiny roar, however the volume got progressively louder as the hole got bigger and when the exhaust eventually dropped off, the sound was deafening, so much so a friend threatened to report me to the council for noise pollution. I eventually took it into an exhaust centre close to where I work and the guy there was delighted. He said he'd heard me driving past for months and wondered when I'd eventually pull in.

The philosophy of leaving things until they really break once got me into more trouble than I'd bargained for and was the catalyst for my inevitably doomed relationship with the AA. My water pump began to leak. The little trickle at first meant no more than the occasional topping up of the radiator, but as the weeks went by it meant that I had to fill it up after every journey and ultimately that I had to carry water and stop to fill it up every twenty minutes.

I started to carry a big red bucket around with me. I would call at friends' houses carrying this big thing and as I was about to leave, nip into the kitchen, fill up and say my goodbyes. This went on for about two months. The point eventually did come when the whole thing snapped and it couldn't have come at a worse time. I was living in London, it was Christmas Eve and I wanted to get home to Yorkshire for the family celebrations. The Maxi at this point could only go for about ten minutes before I was choking on the steam.

I took it in to the local garage, explained a sense of urgency and the guy agreed to fix it if I went to get the water pump, which I duly did. I dropped the car off and he promised that I could pick it up at four o'clock. When I turned up at the agreed hour, the garage was locked with my Maxi inside it. I hung around for about thirty minutes when I saw the mechanic stumbling out of the pub.

'Ah, you are ruining my Christmas man,' he offered.

I offered that he was the one who promised to do it, and he should get it sorted or there might be a little trouble. I was told to come back in two hours. I returned as instructed to find this time the party had moved to the garage, my Maxi untouched. It was now nearly seven o'clock, the buses had stopped running and there were no more trains for four days. It was clear he wasn't going to fix my car, I exploded in rage, threw the water pump in the back and headed off north through London and up to the M1. Having to stop every ten minutes to fill up with water and trying to get through London is not the ideal combination. It took three hours. I finally got to the M1, I pulled over, let the whole thing cool down, filled it up with water and made a mad dash up the first six miles to Scratchwood Services, where at half past ten on Christmas Eve night, I finally abandoned my Maxi.

As you can imagine, trying to hitch a lift at such a time was not easy. The only people on the road were the Scots and they had no intention of pulling in to Scratchwood. It took me all night, I finally had to get my dad to pick me up from Sheffield town centre and I made it home just in time for my mother's Christmas dinner. Food never tasted so good.

At some point between Christmas and New Year, I joined the AA in Pontefract and hitched back down to my car. I called the AA out and the man couldn't have been kinder. I pretended I didn't know what could be wrong with it, he quickly diagnosed the water pump and as it was a long way to relay the car back, he said he would go off to find a new part and fix it for me. It was then I had to come clean and point him in the direction of a spanking new water pump sitting on the back seat. He fitted it and I was off.

The new-found security the AA had given me led to many adventurous journeys and seven call outs in the first six months. All this stopped when a letter arrived on my doormat suggesting that I shouldn't call them anymore. Nowhere in their publicity does it say that there is a limited number of call-outs you can make. You need a Maxi to discover that.

So the Maxi can be an angel, but it can be the devil in disguise. I must say that I have received many brief notes from people saying that the Maxi had never let them down and they could rely upon it to get them to their destination. However, the people with complaints went into much more graphic detail, so the following section is full of such stories. I appreciate it may be a one-sided account of what can be viewed as an essentially solid and trustworthy car, but if we were honest with ourselves, accounts of people's misfortune make a more interesting read than the propoganda that tells us how great things are. If you are sensitive to Maxi criticism, I will attach a health warning to the following section and advise that you skip it.

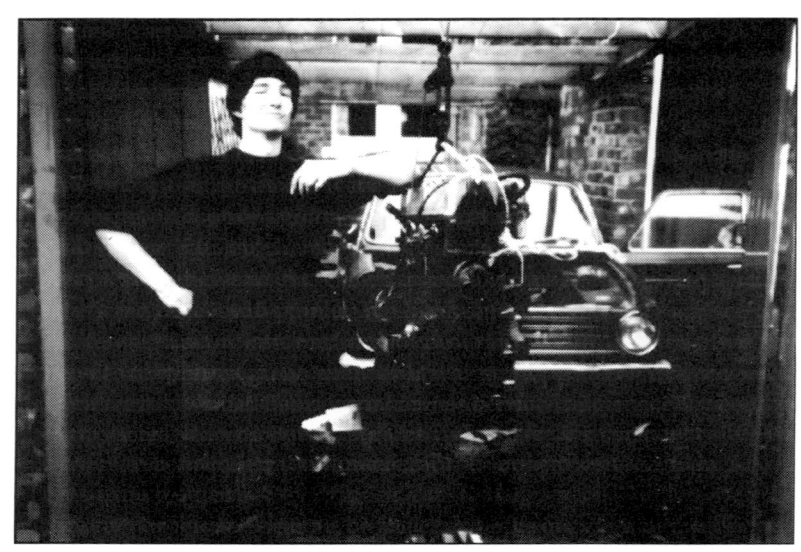

Hang 'em High

The Maxi engine in the above photograph looks disturbingly like a heart and the man, Mark, like some evil butcher.

I often think that mechanics are surgeons who never had the same opportunities in life. I am always amazed when they dive under a bonnet with a spanner sticking out of their back pocket. They mumble, they reach for the spanner, there is a clanking sound, they stand upright and while wiping their hands on a dirty rag say, 'Start it up.'

And when I turn that key and it starts I scream, 'Miracle.' Perhaps if Maxis had been around in biblical times, they might have featured in one of the gospels, a sort of poor-mans Lazarus.

Brake On, Brake Off
Break Up, Breakdown

The Austin Maxi, I remember it well. By the standards of its day, a very sensible piece of family car design. Even by the standards of its day, one of the worst-built vehicles ever to be put on the road.

We must have bought it in about 1971 - a green job, about twelve months old. We had previously owned and been happy with an elderly, decrepit but reliable Mini, and this probably influenced the choice of the Maxi. We knew, of course, it was something of a gamble: the conventional wisdom was that your chances of gearbox failure were about evens - but we liked what we saw, and decided to take the risk. We traded in, I remember, a Morris Oxford estate which was showing the galloping rust which typified four-year old cars of the early seventies for what seemed at the time a more modern design which would do the same job of transporting us, two small kids, a spaniel and all the associated paraphernalia.

And it could do the job very well - when everything was working. It was quite spacious, comfortable and economical. Its best achievement was to take all of us, plus Grandmother and luggage, on holiday to Scotland - pleasant motorway cruising, no problems, and a fuel consumption approaching forty to the gallon, which was more remarkable then than it is now. I recall that it was possible to lay the back seat flat, cushions uppermost, to provide a bed for the children: we tried it once, but they were too amused by the spaniel walking over them to sleep much. The front seats, I think, could be reclined to join up with the flattened back seats to provide an adult-sized bed: a sort of cut-price passion-waggon. We never did get round to trying that, as we always had luggage or family or both cluttering the place up.

It would be nice to remember it as well-designed family transport. Sadly, that isn't the bit that comes readily to mind. I can still pinpoint the exact spot, on my route to work, where I found myself stirring the gear-lever round like a porridge-spirtle to try to locate any surviving gears. The gamble had failed.

Then there was the day we pulled off the M6 and into a local Austin agent's, with the engine missing badly. The foreman quickly diagnosed tracking on a faulty distributor-cap, and with naive optimism I asked him to fit a new one. Shaking his head firmly, he led me to a small heap of defunct distributor-caps, all from Maxis. He said that he had been waiting weeks for replacements. Strangely, Leyland or Rover or whoever owned BMC by the middle of the eighties still hadn't solved the smooth production of this rather basic component: I waited six weeks for one of the last Ambassadors.

The problem of the warning lights was more entertaining, and more readily solved. We were in Exeter on a Saturday morning, when every warning light on the dashboard lit up at once - the whole shooting-match. Not knowing which particular terminal affliction had struck, for it seemed unlikely that all systems had failed simultaneously, even on a Maxi, we rushed in panic to catch the local agent before he closed.

The resigned mechanic didn't even bother to come and look; he just went and collected a spanner and a very small nut and bolt. 'The problem,' he said, 'is that they earth the dashboard harness to a self-tapping screw. After a few thousand miles, the screw shakes loose and falls out. We replace it with a nut and bolt, and you have no more trouble. We've done about a dozen.' As he walked to the car he added, with a tinge of bitterness, 'We keep telling the factory to use a nut and bolt, but they won't listen to us.' He then adopted a most interesting position - upside down on the driver's seat, legs hanging down over the back, head under the dashboard - while he did the job. He was a very nice man, and I don't recall he even bothered to charge us: he just seemed to regard Maxi dashboard harnesses as one of the minor burdens of life.

The tailgate, on the other hand, was a more serious manifestation of sloppy workmanship which defied correction to the very end. The problem was not that when one of the tailgate struts failed it gave a passable imitation of the guillotine, though you might have thought they could ensure that the remaining one at least absorbed some of the weight, but that the tailgate simply didn't fit. Every so often, as you were howling merrily along, there would be a whoosh and a blast of air and instant ventilation as the struts - now working very happily of course - shot the back of the car skywards.

The most serious incident came one night just after we had joined the motorway for a 200-mile journey to relatives in Devon. We pulled into the hard shoulder to check anxiously that the worried spaniel was indeed cowering safely in the back, rather than spinning along the M6, and then hastily transferred her to the security of the back seat along with the children: we reckoned we would lose the luggage rather than her. After we had wrestled the tailgate closed again, and I was about to drive off, a Police Jaguar pulled in ahead to investigate and reversed towards us at high speed. In the pitch dark, watching his reversing lights hurtle towards me, I had the bizarre optical illusion that I was driving fast into the back of him - despite the engine being switched off, and my foot hard down on the brake. It was distinctly unnerving; subconsciously, it may have been the last straw.

We had a very good local garage, and the body-shop man was a gentle, patient veteran who worried away at the offending geometry of tailgate and body-shell to no avail: on one occasion he summed it all up by saying despairingly, 'Look, sir, let's face it - I'm trying to put right something that should have been put right in the factory.'

Which really said it all about the Maxi: the construction was awful. At that time, friends ran a Renault 16, and the Maxi always seemed to me to be the British equivalent of that very successful and popular car. They were both pioneers of the family hatchback; they both offered flexible family transport. The Maxi always seemed to me better looking, with better handling, better ride, better accommodation - generally, a better design. But the Renault worked: it kept running, and it didn't give problems like the Maxi did, and it was properly screwed together.

I recall quite clearly how I came to sell the car: the road-fund licence was due at the end of the first year, and I simply wasn't prepared to pay out for another year's tax. I've never felt that way about any car, before or since - which in itself gives the Maxi a unique distinction among all the vehicles I have owned, and imprints it in the memory.

I'm not sure why I went to try the Saab - a white, two-door 96 saloon which was not at all what we really needed. It was love at first drive - a taut, rattle-free car that felt right for me within the first mile, and began a devotion to the marque which only ended when the 900 series took it out of the range my pocket. There were advanced safety features which the Maxi had lacked, though it was the superior performance and handling which made me realise what a tub I had been driving, and above all there was a feeling of solidity and strength and sound engineering - all this in a basic design about two decades older than the Maxi. The Saab 96 - there's a car to really celebrate; I still can't pass one without wanting to stop it and ask the owner if he wants to sell. I don't feel that way about Maxis somehow, not that I see many of them these days!

I bought the Maxi with my flatmate from 'Dodger' in Hackney for £200. It was ten years old, had four new tyres, a year's MOT, the sills weren't dry from being newly black painted, and as my mate Steve drove it home in the traffic on the Lea Bridge Road, we discovered the clutch was slipping.

Steve had just had his car written off through a tail end shunt. Since I cycled to work I had felt obliged to let him use my MG each day. After a week of lending my cherished car, I was jumpy and nervous at work, so I suggested we buy something cheap as a runabout, between us. Hence the Maxi.

Being a rather naive enthusiast of cars, the first thing I did was to change the oil and filter. Whilst I was commending the smooth tick-over to my dad, the car was slowly but steadily leaking out the new oil - which soon dad was standing in. I had fitted the wrong filter.

Steve and I shared the car for about six months, in that time we had a new battery, alternator and clutch but ran out of petrol about half a dozen times, each of us showing a curious reluctance to feed it with fuel. In the end we took to carrying spare petrol in a can in readiness for the inevitable.

The first time the garage fitted a new clutch the oil seal leaked between the crankshaft and the clutch, so they fitted another. This was in fact an ominous sign of things to come. But the great thing about living in Walthamstow was that every street corner that didn't have a pub, had a back street garage instead. Competition was rife, prices keen, promises honoured.

Although I had an MG, I had otherwise never rated Leyland cars, but this car had virtues. It started every morning right through the winter, it was tough, it was absolutely cavernous, and with a street value of £200 you had no fear leaving it anywhere in London; it was always there when you came back. My girlfriend Chris - now my wife - had just bought a brand new Sierra - hot off the press, but at £6000 I proudly proclaimed the same money could have bought a street full of Maxis - thirty in fact!

Steve bought a Chrysler Horizon and left his half of the car to me. The Maxi still ran reliably, but the engine was worn out, in fact the car wouldn't do much over 40 miles an hour and it drank so much oil I took to scrounging friends' dirty oil and just topping the engine up. So I decided to fit a new engine - with the help of someone who 'knew about cars'! But when I say new engine, I mean a recon short engine from a Lewisham scrap yard.

Fitting took a weekend, and wasn't without problems. Since that old clutch oil seal had leaked again, giving the car that automatic feel, we fitted another clutch and seal while we were at it. Buying a short engine of course meant a tremendous amount of extra work: taking the top end off, separating the block from the gearbox and replacing that dodgy old crankshaft; then reassembling the whole thing with new seals and gaskets. Much easier to buy the whole lump in one go and just drop it in.

Proud of my weekend's work I suggested to Chris a running-in trip to Southend. We got as far as Ongar, and on went the oil light - yes, that seal was leaking again, and very badly. Time for one of Walthamstow's back street garages to come to the rescue.

I decided to adopt some 'rules'. MOT every six months and never spend over £50 on it in any one month. I didn't want to get into the trap of throwing good money after bad. I kept going like this for two years and 30,000 miles and had some fun on the way. The greatest source of amusement was always the clutch and the gearchange. I used to keep a few split pins in the ashtray, because the pin on the linkage from the gear lever to the transmission had broken and every now and then my improvised replacement would break. The gear lever was so sloppy you didn't always know immediately it had happened. I remember once rolling up to a set of red traffic lights, leaping out and under the front of the car with split pin and pliers, reconnecting the linkage, and back in the car again before the lights had changed.

The clutch was the other problem. The bolts on the clutch housing were so cross-threaded that when the clutch pedal was pushed the housing moved out as much as the clutch plate moved in. I cured that with a bit of wood wedged between the housing and the bodywork.

Then on a weekend away the clutch release bearing packed up and I drove the whole weekend making clutchless yet silent gearchanges; because everything was so worn it was really quite easy. Only parking was really difficult, because you had to stop and start the engine each time you wanted to change direction.

The car did deteriorate and the new engine started drinking almost as much oil as the old one had. But Chris and I took it on one last epic journey to Brittany, on the probably misguided basis that if the car folded we would abandon it and catch the bus home. The car covered its last 1000 miles with us in Brittany with a bowling wheel bearing and a curious electrical fault which allowed the car to start brightly from cold, but if you drove down the road and stopped for petrol, the battery would not have the power to turn the engine over. Easy to push start again though.

When we got home I had my sights on a new car. For £895 I bought a one owner Maxi 1750HL with twin carbs. What I couldn't believe was, the salesman apologised for only being able to give me £200 part exchange on the old car! That's 100% profit after two years cheap, fun-packed motoring.

That was nine years ago, I still have the MG, but all I have left of the Maxi is a can of damask red spray paint, a Haynes manual, and the memories. By the way, did anyone else keep a broom handle in the boot to hold the hatch open?

There have been many incidents with the Maxi involving mechanical problems, many of them in fortuitous circumstances. My wife was taking my daughter down to Middlesex for a college interview. She was travelling down the M1 and as it joined the M6 she found herself in the fifth lane with a puncture. She reached the hardshoulder after what seemed an eternity and took off a hub-cap that rolled down the motorway edge like a bit of tin foil. As she struggled to loosen off the wheel nuts a man strolled up and asked if he could help. What luck to have pulled in next to a mechanic repairing a lorry!

When my son was attending college interviews, my wife took him down to Brunel University and stayed overnight at her brother's in Maidenhead. The Maxi appeared to be lazy starting when setting off from Maidenhead and again when leaving Brunel. Parked up again at Streatham at my daughter's digs, the Maxi started well when leaving, and after filling up with petrol also after a break at Leicester Forest on the way to Hull. When I wanted to start the Maxi the next day it had given up the ghost - the starter motor had gone - but the 'little pet' had got my family home!

The Maxi went 'round the clock' on our way from Hull to a funeral at Chichester. We had timed the journey very narrowly and on the way to the crematorium met the funeral cortege at a roundabout. Luckily we entered the roundabout first but then typically manoeuvred our way out of good fortune as we made a wrong turn at some traffic lights further on. We stopped and asked an old man the way. He seemed to speak so slowly that we felt the funeral would be over before he had given us directions. He ended by saying that he hoped it would be a long time before he was going to the crematorium!

As we came to the traffic lights again waiting to turn left the funeral cars were waiting to our right ready to go straight on. Luck was with us again as our lights turned green first. We were able to park up and join the party as if we had time to spare.

The Maxi has had numerous bumps that have hardly affected the tank-like body work. It just looks like an old soldier who is all the more interesting for collecting his battle scars that only add character. I have owned the Maxi for eight years now and I think I would rather sell my wife than my car!

My Austin Maxi was a white, 1750 cc model, reg MGN 814V. It was a company car, I had no choice of make or model and took delivery, from new, on 10 October 1979. I handed it back to the company fleet manager on 3 March 1982, having clocked 80,931 miles. It had been 28 months and 3 weeks of sheer frustration.

The Maxi was, without doubt, the worst quality car I have driven in over three quarters of a million miles' driving.

While still within the warranty period the rear wheel bearings were replaced about six times - sometimes with part of the rear suspension as well. Two gearboxes were fitted in the first couple of months, front wheel bearings also gave trouble, the speedometer drive failed on at least two occasions, the radio caught fire and - on a memorable occasion - the left hand side suspension went 'flat' leaving me fifty miles from home with no springing on that side.

From first to last it leaked water like a sieve, despite much rectification work. Not for me the snap decision to go through the car wash. My Maxi first demanded all windows have chamois leathers trapped in them to stop the water pouring through the alleged 'seals'. Even then one had to sit in the car with a large cloth to field the inflow from various gaps.

As it was a company vehicle I at least got a hire car during the Maxi's frequent absences from the highway. I made sure these were Cavaliers, Cortinas or Solaras. Having previously driven fifty-six thousand miles in a Marina estate - again not my own choice - I had no faith in Leyland products.

In the warranty period I met many Leyland service engineers. Time and again I demanded to know if they thought the car suitable for business use of up to a thousand miles a week. Their answer was always the same, 'It's a good family car, sir.'

I lived in Coventry then and once met a Leyland manager at a party. I started telling him my troubles but he was a bit aggressive. 'It's the way you drive it,' he said, 'you have no sympathy with the car.' He didn't actually apologise but he certainly went a bit quiet when I politely asked how my driving could cause the dipstick to snap in situ and the locking petrol cap to disintegrate.

There was one positive feature. One winter afternoon I was hurrying home to Coventry from London. It was snowing hard and everyone was crawling on the two inner lanes of the Ml, leaving the outer lane with a virgin covering of up to six inches of snow. Brimming with confidence I pulled into the outer lane and put my foot down. Inner lane speeds were about 35 mph, I was soon clocking 60 mph. The Maxi was rock solid, no

Destiny

The top photo shows what happens when a Maxi has been a bit naughty. The bottom one is a Maxi that has been very, very naughty.

twitching, no wandering, no hesitation. My only fear was that some loony would pull out ahead of me without looking in his mirror. All was well, soon I had a 'train' of followers in the outside lane, though none kept up with me.

Once home I parked, went indoors and forgot the car. On starting it next morning, I couldn't select any gear. Neither could I turn the steering wheel or release the handbrake.

Looking underneath soon revealed the problem. Compacted snow from the previous day had completely frozen the underside of the car - encasing the exposed gear change and handbrake mechanisms in ice. The front wings were also solid with ice. I won't name the tools used to remove it but they were more likely to be found in a garden shed than a workshop!

Apart from the snow performance I have nothing good to say about the Maxi. Windscreens leaked and shattered, caused - as I understand - by flexing of the bodyshell. Window winders failed. Engine and gearbox oils dripped out. For a five speed car - my first - it was not very economical. Performance was mediocre by the standards of the time.

The saving grace was that Leyland, and later my company, met the bills for repairs or hire cars. As a private owner it would have broken my bank. I averaged about 700 miles a week for nearly two and a half years and never felt confident I'd reach my destination. I wish I'd kept a diary of the faults which developed, those mentioned are just some which come immediately to mind.

Viewed in isolation my Maxi could have been written off as a rogue vehicle. However, other people I have spoken to on the subject claim similar experiences - in fact I've never heard of a satisfied Maxi owner. One sometimes hears people say 'They don't make them like they used to.' Thank God.

Like so many British-made cars, the Maxi was brilliant in concept and design, but generally poor in execution and finish. I shall always cherish the memory of it as the first front wheel drive, first 5-speed gearbox, first hatchback general purpose car we knew and owned, and I remember it vividly as a true family car. It belongs to this brief story of the Sommershield Maxi, that neither I, nor my son, have had car accidents since those which we had with the Maxi in the 70s.

We bought it second-hand in April 1973, when our old Rover 90, brought with us from England when we emigrated in 1970, was condemned by the Danish authorities as part of an MOT test. It never took to the Danish habit of pouring liberal amounts of salt on wintry roads, so rust and corrosion were chronic.

The Maxi was white and looked absolutely splendid when we first saw it in the dealer's window; our oldest son, at the time at Sussex University in Falmer, warned us that the Maxi was generally considered to be the ugliest British-made car ever, but I never shared that view. It was low-slung and squat, but long in body, and with an enormous amount of space inside which was just the job for a family with four teenage children plus a dog. We used it for going out in party togs on Saturday night, and then carting horse manure in plastic sacks home to the garden on Sunday morning.

The 1750cc engine was all right, no more than that, but it was reliable enough though it liked its oil, plenty of it. It was the first car we had had with front wheel drive and a 5-speed gearbox, and what a boon that was! In fact, the only snag in the machinery department was that the gear change mechanism was of the original 'rope-and-pulley' type, and the gear lever felt rather like a stick of liquorice.

The first accident happened in the summer of 73, when I was touring in Norway. On a narrow mountain road, very touristy, between Grotli and Geiranger in the West Country, we came round a bend and were confronted by a Volvo which we couldn't see before it was too late. Bang. Smash. Head on. End of that trip as far as the Maxi was concerned; what a blessing a fully comprehensive insurance can be!

The second time I realised how horribly vulnerable the Maxi front end was, by virtue of the whole engine and transmission assembly sitting on a hinged subframe which very easily became unhinged. This was about a year later, when on my way to work in Copenhagen the bloke in front suddenly jammed his brakes on. After the normal reaction time, so did I, but his brakes were much better than the Maxi's, and so another expensive meeting took place. And since the bloke-in-front's car happened to be a VW Beetle, which is as vulnerable at the back as the Maxi was in front, it was a very expensive meeting indeed. What a blessing a fully comprehensive insurance can be!

With the engine-and-transmission assembly sitting on subframe number three, we now motored quite happily for a couple of years. The end came in January 1978, when one of the boys drove the car on a real winter's day, on a real winding country lane. The surface was icy, and in a bend the car got into a skid and the front end rammed a wooden power line pole. And this time it wasn't just the subframe but also its support points under the body, plus a lot more, which copped it. So the insurance company, who must have realised by then that the car was jinxed, said, enough! and bought themselves free with an amount equal to the average value of a 7-year old Maxi. Which didn't buy a new Maxi, of course, and as I remember it, the Maxi was obsolete by then anyway, and we bought something far more orthodox and boring.

The Maxi was an idiosyncratic car as one never knew what tricks it would next discover. The Maxi was allocated to me as a company car many years ago. I could have chosen a Ford, but for some weird reason I opted for the Maxi. Although my company car intake was 15 in all, the Maxi and its habits has stuck in the memory most.

The first Maxi was a grey colour job, one of the early models which had a cable gear change. A bad design. Mine stretched 50 miles from home and it was impossible to engage first, second or third gear so I came home with clutch slipping and much revving at traffic lights.

The next Maxi was a maroon job, the registration letters were FKL, the car was soon dubbed 'Fickle Firkle', that being a catch phrase from a TV show. The company for whom I worked used to run cars up to 50,000 miles, which is not a lot by today's standards, and it came to pass that Maxi FKL began to develop funny problems as the mileage increased. The hydroelastic suspension system developed gremlins and overnight the car developed either lists fore and aft, or Port and Starboard. Whilst driving through the Dartford Tunnel one freezing cold morning, the windshield suddenly fogged up, steam arose from the driving well and my shoes, trousers, and socks were deluged with hot water. With a 20 ton lorry sitting on my tail, I just had to carry on, feeling like a lobster being boiled feet upwards. The hose pipe which ran behind the bulkhead had split.

Another memorable instance was on the A2 approaching the Dartford Tunnel, a dark autumn night. Overtaking another vehicle I moved the traffic indicator, there was a blinding flash, all lights were doused and I had to pull over rapidly to a type of hard shoulder. The whole light/ indicator assembly had just fallen apart. This meant a mile walk to a garage and as it was clocking off time, not a lot of assistance but I managed to hire a car and eventually arrive home.

And then there was the alternator! Why do these things happen in Kent and on the A2? A murky, foul, drizzly night approaching the Dartford Tunnel, there seemed to be a lack of vigour in the lights, windscreen wipers and engine performance. The problem became a reality in the

Tunnel, I just cleared the exit on the Essex side and the car just died. That meant a long walk to find a phone to ring my wife and the AA - who put a booster charge into the battery to allow me to travel just a few miles to home.

I think the Maxi has some kind of objection to the Dartford Tunnel.

The Tank's engine was always wonderful - it started first time every time - but the bodywork by the end of his life left a lot to be desired. About two years before his demise the rain started getting in. It had quite a few holes so it was never clear which particular hole was the culprit. It used to gather behind the dashboard and slosh about alarmingly when I turned corners - any sharp bends or abrupt stops were followed by a cascade of water from somewhere behind the steering column. A few days of solid rain would mean puddles on the car floor. After a while I was so worried about the effect on my passengers that I offered them carrier bags and we would proceed down the road in the Tank with our knees swathed in plastic, bravely ignoring the burning smell from the fan. It really would have been drier to walk. In fact, it was the water that eventually did the car - just about everything rusted and gave out, including the driver window which fell still in one piece into the door cavity. I was in the car park at work at the time and luckily for me one of the chaps took pity on me and wedged the window closed. I spent the Tank's last weeks leaping out of the car to collect car park tickets from barrier machines instead of slowly winding down the window.

The only other fault worth mentioning, was the front passenger side wheel. In the space of ten years no garage the length of Britain could explain why it was prone to tyre punctures. One year we toured Scotland for two weeks and had four punctures in the same wheel, one of which was discovered by a police woman as we were about to drive onto the Skye Ferry. To the end of the Tank's life. It was a total mystery.

I was once told by an RAC mechanic that cylinder head gaskets in Austin Maxis were 'very suspect' - other drivers will no doubt be able to confirm or deny this. This was after he'd towed me off the M6 at Birmingham. I'd looked in the mirror and casually remarked to my mother that something behind was on fire. I then realised that all other vehicles were sensibly keeping their distance from the thick black smoke that was belching out of the Tank's exhaust.

It makes it sound as if I was driving a death trap about for years, but the Tank actually charged on and on and passed MOTs at perfectly respectable garages. It was when it got to the stage that it was me and not the car that was rusting that I finally decided that the bodywork had definitely had it. I have been in no other car that was as nice to drive or as comfortable to sit in and certainly no other car with as much character. They definitely don't make them like that any more.

I was talked into buying my Maxi. I had, for many years, been passionately addicted to Triumph 2000s, but my last one, a white 2.5 PI estate, was becoming elderly and costly and I was urged to be 'sensible'. I'm sensible about many things, but cars have never been one of them; I've always liked something a bit out of the ordinary! However, my boss and a British Leyland salesman persuaded me to invest my county council car loan in a metallic blue Maxi, which I called Woofu, on account of its registration WFU.

I agreed that the car was roomy and practical and even economical. I had three growing children, lots of luggage to transport and a job which involved moving wheelchairs and aids for disabled people. Why then did I sit in a traffic jam with tears streaming down my face on the first day I owned it? My mother, a passenger, thought I'd taken leave of my senses,

'I think it's a lovely car; you've never owned a new car before. What are you being so silly for?'

Well, I'd been bereaved. I saw a Triumph ahead of me, and admiring its lovely sleek lines I felt so boring, ordinary and utterly 'sensible' in my box on wheels. It seemed like the onset of middle age, and I was only in my 30s! Also I was shocked that my mother actually approved of one of my decisions - that got Maxi off to a bad start!

It paid me back though by turning temperamental and refusing to start on quite several mornings. Fortunately I didn't live very far from the Leyland garage. Maxi wasn't cowed though. He saved his best tantrum for a far more inconvenient moment.

I was taking my two sons back to boarding school - on the last moment as usual - and laden down with trunks, tuck boxes, skateboards and radios when in a queue of traffic under Skipton railway bridge he stopped dead. No gearbox. Quite spectacular really and certainly a stylish 'sulk'. We pushed him to the side of the road in front of an interested audience and I rang the AA. Some time later my sons were deposited at school by a yellow AA van, which somewhat lowered their status in the eyes of their peers. I, meanwhile, settled down, bookless, to a long wait for the breakdown truck and arrived back in Lowth at 2 am next day.

I think after that Woofu and I must have declared a truce, because I remember many subsequent miles of trouble-free motoring, and I must have kept him quite a long time as my daughter passed her driving test in him when she was 17. Filled with a curious mixture of emotions, I stood and waved her off on her first solo journey, but that's another story.

Buying a new car always seems to be the result of a process of adjustment. We had no plans to get rid of our Hillman Minx until a friend's brother died. He was alleged to have an almost new estate car. I was on holiday in France at the time but I was telephoned with the news that we were being offered the first option. I managed to get hold of some literature and identified a hatchback as the possible model. It seemed a bargain. I spent several sleepless nights thinking about it. When I got home, we rushed off to see this marvel and found that it was not a hatchback at all but a rather odd-looking down-market model. We disentangled ourselves from this situation but were now sold on acquiring a newer car.

It was not long before our garage produced the Maxi, 18 months old, but with a slightly dodgy gearbox. I managed to change down into third on the test run and thereafter survived reasonably well, although it was possible to lose track of which gear one was in. I always found my ability to double-declutch came in handy, as it still does with a Metro. I seem to remember losing the use of one gear once, but at least it was a forward gear. We once lost reverse on one of our Minxes and had to drive in circles.

One disconcerting fault the car developed in the last year of its life was an inability to go fast. I would set off and do a couple of miles and then it would slow down just when I had entered the M25 and crawl along, however much acceleration was tried. The fault was intermittent and did not show up when the car was being serviced. It was very embarrassing, crawling along the M25, and a trip to Reading was a nightmare. Eventually a mechanic came to collect the car for an investigation and it really went to town on him - he had to send for a tow. The brakes were seizing up.

One convenient breakdown was on a holiday in Yorkshire. We had just done our usual visit to Reeth to eat ice-cream on the village green, and to admire a beautiful, tomato-soup Maxi in a local yard, and were travelling back to our holiday cottage in Thirsk, when things went wrong. There had been some funny knocking noises when doing a slalom through country lanes and now it came to a head. We crawled to the nearest garage in Richmond and were waved on to the local Austin dealer. There a failed half-shaft was diagnosed. It was late afternoon but my poor mum sat in the middle of the garage between the cars while a repair was made with a cannibalised half-shaft and, late but relieved, we were able to make it home.

I never tried sleeping in it but I did sit cross-legged in the back, indexing an old newspaper scrapbook, on wet days in Cornwall.

Front ball joints on Maxis were prone to failure but could be changed quickly, the thing I couldn't do was reset the tracking of the car. This was done by a local garage but always left the spokes of the steering wheel set to the left or right of centre.

Most annoying on a straight road! But no problem; on the way home I would remove the nut on the steering wheel and whilst travelling at 30-40mph in a straight line remove the steering wheel and replace it on its splines in the right position. Easy eh!

Oil consumption - valve guides I believe - became so bad that it was normal to purchase a gallon of oil with the weekly shopping.

After a few years the hydroelastic suspension sagged. Being unable to afford to have it pumped up properly but having access to high pressure pumps I decided to do it myself. I used a mix of antifreeze and quickly returned the car to its current trim height. All good until the following bitterly cold winter when I discovered I hadn't used a strong enough antifreeze mix. The first few miles each morning were accompanied by a sloshing sound of ice in the pipe work under the car and a fairly stiff ride.

Part of the 'fun' of winter driving was to be had on fresh snow-covered roads across the moors in North Wales. The Maxi was so stable that you could jam the handbrakes on and see how many times the car would spin before stopping. The beauty of the Maxi was that if the handbrake was released whilst spinning it would straighten up right away. Much laughter! I often wondered what following cars made of the tracks in the snow. This same trick could be done in the summer on grass and at high speed too, the drawback was when you came to a halt the inside of the car would be full of grass cuttings. Even more laughter!

The Maxi is a sound car but it does have mechanical Achilles heels which are three in number; ground clearance, loss of oil and the ring gear.

You could pump up the suspension so it looked like a pond-skating insect, but it bevelled the tyres. At normal height, I knocked a piece out of the sump on a folded down post in a car park, and the car bled to death on the spot. Embarrassing, since it was a private car park, and I had no means of clearing it up, nor moving the car until I had summoned assistance. With engine and gearbox sharing oil, this was a major repair, and the insurance company eventually treated it as an impact accident with wear and tear features. Hundreds of pounds.

I learned to park outside strangers' houses rather than friends'. Apart from the black deposit of oil when stationary for only a few seconds, it was possible to gauge the speed of approach and rate of deceleration from the interval between the black spots along the road. I could never park on anyone's driveway - I ruined my own - but learned eventually that the cure was never to put any oil in. It didn't make the slightest difference. Austin should have sold an air-cooled version without changing the design at all.

Another gimmick was that an overfilled petrol tank on a hot day would relieve itself by expansion and dissolve the Tarmac in the staff car park. And why did the plastic chrome bits go brown?

The ring gear. My tame mechanic, who had been very successful with my Escort Estate even though he never coaxed it above 62 mph, suggested that the oil leak was to do with the seal around the ring gear. Maxi drivers may have noticed that the ring gear transfers power from the engine to the gearbox below it, and is a composite of about three different metals: steel, chrome and bronze. Nearly 20 years ago it cost *eighty pounds* for the part alone. Major engineering was required to replace it, and the problem was exactly the same afterwards. Though the car is long gone, my curious but expensive door-stop-cum-paper-weight is still a talking point.

I liked the colour, the front wheel-drive, I never skidded it, the 'big car' clunk when the doors shut and the rearward visibility for parking, as well as forward visibility for getting through gaps. In terms of enjoyment, and also for cheap motoring overall for four years, it was the best car I ever had, and I have never regretted the experience.

I am an exclusive Do It Yourselfer. As you can imagine I have experienced most of the disasters and mechanical 'nasties' that can befall these cars. Only yesterday I had to remove the clutch master cylinder - have you ever tried it? My God, what a design! The flange of this cylinder ledges behind the brake servo with an overlap of about an eighth of an inch - and for no apparent reason it is mounted on studs so that it will not come off. About 2 o'clock yesterday afternoon I would have been delighted to assassinate slowly and painfully the draughtsman who failed to notice this. There are two solutions to the problem; remove the servo, apparently not an easy task, or resort to the most appalling violence and break a tiny and redundant piece from the master cylinder casting. I chose the latter course and the car is now back in service.

There are in my experience four other weaknesses of this car apart from creeping body rot. They are; the system for locating the front 'hub' on the stub axle with that cone-faced split washer. A nice idea badly executed, easily improved. The clutch. Easy to replace. It would be interesting to know the record time for doing this job. I have done it complete in about 2 hours. Rear wheel bearings. Ball bearings! Not quite up to the job. Why not tapered rollers like the Allegro? Butterfly bushes - nothing wrong with the bushes, just with the way people install them, and the price of new ones.

Of course I have encountered many other mechanical failures but not on a regular basis. As far as the real machinery is concerned I am amazed by the reliability of the car. In all those cars and all those thousands of miles I have never removed a cylinder head and I have never seen inside the gearbox. I have not owned one of the very earliest cars with the legendary cable gear change. It did not seem to last long and yet its reputation goes on - and on - and on.

I have had the front subframe collapse on one car so that the wheels were toed-out by about four inches. Petrol tanks have a nasty habit of rusting through just above the seam, by the left back wheel. I have actually had to change a front hydragas pot - that was quite a lot of welding - particularly of the inner sill near the front. Why it should go wrong there I cannot imagine.

I have slept in the car overnight with my wife in real comfort. I have towed a small caravan for thousands and thousands of miles with it, I am very pleased with it. I don't know what I'm going to do when there are none left. A while ago I bought an Ambassador as a possible successor. It is about 2ft longer than the Maxi but with less room inside. Because of its front overhang I kept bumping into walls and car park barriers so I sold it and found yet another Maxi.

'Maxi baby' proved to be very reliable and only broke down once when she stripped the gears on her inside drive shaft. But because of her

Tardis-like interior we were always able to carry a full workshop of tools and many spares. My mate set off to Maxi Man's, someone I got spares off, while I started work. An hour or so later we were again mobile. Ok, we did have front bearings go on us at regular intervals and the 'pump to carb' petrol pipe kept on splitting, she was fine otherwise!

Once five of us, and I'm sure an illegal amount of camping gear and beer, set off in the Maxi to the Isle of Wight. On landing the Maxi chucked one of her wipers in the hedge, and the wheel bearing had to be re-pinned and tightened. The isle itself is full of Maxis and ours proved well on all the nasty humpy hills on the island, though she did rip off the exhaust on one such hill, but of course we fixed it, no problem!

During her year with me, she was also used many times to tow-start and tow home many cars which she pulled with ease, once pulling the belly off my brother's Skoda. Don't get us wrong, we were basically boy racers at the time, we were just boy racers with a car that could work, not just wheel spin. Many Escorts were put to shame by my brown Maxi.

In her last month in my service we pranked with a nasty old Escort, well not old, new, very new! The driver of the Escort, laughed at our demolished wing but we laughed at the ripples up the side of his car and the gap now visable between his boot and the wing, for unknown to him and many, the Maxi is well built under those wings!

Soon I beat the wing out, put her through the MOT and sold her for £220. I hope she is still being appreciated as before. I missed the Maxi when I bought my V8 SDI because although faster, more comfy and nastier to Escorts, you just can't put 5 adults and a motorbike in her no matter how hard you try. The Maxi is one of my favourite cars.

I suppose I should have known better, but there it stood forlornly at the back of the dealer's lot - Harvest Gold, 'M' reg, 50,000 miles on the clock, its slightly crumpled front bumper added to its dejected appearance, but it was a Maxi and I hoped, the end of a fruitless chase after models in the small ads.

The gearbox failed on the test drive. 'No problem' said the dealer, 'we'll fix that.' They did but it retained irritating vibrations at idling speed that indicated something out of balance. I suggested the flywheel maybe? They checked and things seemed a little better 'They all do this,' said the foreman. I bought it hoping things would get better.

During the two years I kept it, I replaced all the engine mountings, the steady bar rubbers, checked ignition timing, valve timing, oil pressure, cylinder compressions in an attempt to stop the vibrations. In the end I learned to live with it, except that I had to apologise to any new passengers.

To relieve the monotony of working under the bonnet, I replaced a front wheel bearing, the flexible exhaust coupling, had the driver's seat frame welded as well as the handbrake mounting which had come away from the floor, and topped up with minor items like repairing the courtesy light switch, replacing the petrol gauge transmitter, and then the rot began to show in the sills.

All this was forgotten when the 'squeak' began, rhythmically to every revolution of the front off-side wheel. Replace brake-pads - 'squeak'- clean off scale from brake disc - squeak - check everything possible until a friend and neighbour stricken with deafness in his offside ear walked alongside and pronounced, thanks to his monaural hearing, that the noise came from the tyre. At some time the front tyres had been replaced with unbraced radials - the scuffing that followed had led to the inside of the rims having an almost chromium plated look - no wonder it was better in the rain - I couldn't blame BL for this one.

Things went well for a few weeks, then whilst on holiday in Brittany 'le tweet' began. Having unloaded at our site we were happily touring the area when the irritating noise came from the rear - stop - move remaining rubbish around, no 'tweet' a few miles - 'tweet' - repeat process on and off for the whole of the holiday only to find that when we were fully loaded for our return - no tweet. Back home after several unsuccessful attempts to trace it, the same helpful neighbour lay in the back and turned his good ear in every direction - 'It's not from the floor, it's from the roof; he claimed - it was certainly, a quick squirt of WD40 on a dry tailgate hinge and the tweet was banished.

By way of recompense I gave this same helpful neighbour a lift into Birmingham for an appointment with an ear specialist - the gear box linkage came away before we had covered 200 yards, he was late and not very well pleased!

We parted company in Somerset when the clutch gave way. I saw it some time later parked in Birmingham City Centre - I recognised it by the crumpled bumper. I never got round to replacing that. I don't think it knew me. I thought of hanging around until the new owner returned so that we could compare notes. In the event I decided against it, there just wouldn't have been time!

Bonding

People say that it's a sin,
To fall in love.
But I don't care what anybody says
I love my Austin,
Maxi.

It was late one evening in the kitchen of my flat in London that I realised I had bonded with the Maxi. I was teaching myself how to play the guitar and after mastering the basics I was all set for writing songs. I struck lucky this particular night with a lovely, sad sounding three-chord song. I played it over and over and started humming some lyrics; it was around the time I'd just lost my first Maxi and the sadness of the event was still in my mind so the words just flowed out. *The Longbridge Love Affair*, a very silly love song to the Maxi, was born. I didn't know at the time that the Maxi was in fact made, not at Longbridge as the song suggests, but at Cowley.

My first Maxi died in terrible circumstances. I'd been working at the Wimbledon Tennis Championships. It was a very demanding job, starting at 7am I'd work straight through until 9pm for the whole fortnight.

Each night after work it was customary for the staff to retire to The Castle pub in Wimbledon Village. On the third day of the championship I drove my Maxi up to the pub and just as I was parking something snapped. It would rev, but it wouldn't go anywhere.

For once I couldn't just abandon the car and sort out the problem later. I was in a very expensive 'Tow-Away' zone and if the authorities took it away it would have been a write-off, the cost of recovering it higher than the cost of buying a new Maxi.

I left it overnight, but the next morning before work I made a detour to one of the huge car parks surrounding the tennis complex which were staffed by AA men with a tow truck. I explained to them what had happened and I wasn't a member, but seeing that they were sat on their backsides for two weeks twiddling their thumbs, could they see their way to recovering my Maxi and try to fix it. I promised a back hander and they agreed.

I went back to see them during my lunch-break and they told me they'd recovered my car and diagnosed a disintegrated clutch and its housing. I said fix it, they said they didn't have the parts but if I could get them then they would fit them.

I hardly had time to breath during Wimbledon. I usually lost a stone in weight even though I ate almost everything I saw and drunk enough milk to keep a herd of cows in business for several generations.

'The parts are too expensive to buy new, it wouldn't be worth your while, you should get them from a scrapper' they informed me as though I had nothing better to do. But I need the car fixed before the end of the tournament, so I said I'd see what I could do.

Back in the complex I managed to find a *Yellow Pages* and I rang up a local scrap yard and by pure chance they had the parts. There was one problem however, the parts were still attached to a Maxi and the Maxi was on top of a huge pile of cars at the bottom of their yard. I then found myself in the ludicrous position of negotiating with a scrap dealer. By the end of the conversation I had pledged a bigger chunk of money than I was planning on, but if I could get down to their yard the next day I could pick up my parts.

The next day came. I got my workmates to cover for me, I dashed out of the complex, fought for a cab, got to the scrap yard, sneaked past the scrap yard dog, got the parts, got back to the Wimbledon car-park and dropped the clutch system onto the lap of a very impressed AA man.

'You're crazy,' he offered.

'Fix it.'

And so it was done. Finishing Wimbledon is always a great feeling, but that year it was extra special. I'd snatched my Maxi from the jaws of death against all odds, it was a backhand volley of a salvage and I saw chalkdust.

The AA men gathered round and cheered me off as I picked up the car.

'You're mad' they laughed as they waved their hands and my £100 at the Maxi as it made a more dramatic exit from Wimbledon than John McEnroe ever has. On the Centre Court everyone was cheering Stefan Edburg, but in the car park out the back everyone was cheering me and my Maxi, and it felt good.

From euphoria to desperation in twenty four hours. The day after my glorious exit from Wimbledon, tragedy struck. I was giving a friend a lift in Streatham when something else snapped. All indicators pointed to the gearbox. We pushed it for half a mile to the nearest backstreet garage, straight opposite Streatham Common Station. The guy there said it probably was the gearbox but he'd have to charge me £150 just to look at it. I said I'd sleep on it. I was at the end of my tether, I left the car outside his garage overnight and for the next three months. I did go past it many times. The boot on it never locked, and after a couple of weeks the mechanic had got wise to this and started using it as a storage depot, there were empty boxes and paint tins all over the place. When the tax eventually ran out the council towed it away. It was the end of an era.

I didn't like being without a car for too long, so the hunt was on. I bought a copy of *Loot*, London's advertising paper and went through the motors section. I wasn't particularly looking for a Maxi but I saw one. It was in Epsom, so I grabbed a friend and got a lift down. It was beautiful, a bright red one. I bought it for £125. As I drove it back to London the memory of my old Maxi washed away, out with the old, in with the new. At this point I knew I was hooked. It became clear there was no point in buying any other kind of car.

The red Maxi lasted about ten months. I'd left London and moved back to my home town of Featherstone. At the time there were four Maxis in the town, one man owning two. I set myself a challenge to buy them all up. When the MOT was running out on the red one, I noticed one of the Featherstone Maxis was for sale. It was up for £400. I couldn't afford that amount of money, but I went to see it and put in an offer of two hundred pounds. He wouldn't sell.

A few months went by and I had to get rid of my Maxi, the cost of putting it through the MOT was prohibitive, so I advertised it for sale. It went on the first day for £65. Ever since I'd had my two hundred pound offer turned down, I kept seeing the Maxi reappear in the newspaper, and he was bringing the price at a rate of £25 per week until it eventually reached £195. I called back at his house, said I was still interested in the car but there had been a change in my fortunes and I could only offer him £100. He snapped my hand off.

So in terms of my quest to clean up Featherstone's Maxis, it was one down, three to go. This new one was maroon, it was in great condition, he had looked after it. It was a very strong car and lasted a whole year.

I had been to Wolverhampton to an annual reunion dinner with some old friends and was late in setting off back. I really pushed the Maxi, I was going 90mph all the way up the M1. On one of the downhill stretches I really opened it up and was bombing down the fast lane. I glanced across at the faces of the people I was zooming past to see expressions of horror. I was thinking they couldn't deal with being blasted by an aging Maxi, but I when I got to the bottom of the hill and glanced in the rear view mirror to see a thick cloud of black smoke stretching as far back as I could see, I realised what they were worried about. I relaxed the throttle immediately and thought I'd got away with it. Twenty miles further down the road and the Maxi engine rolled over and died. I managed to get it home and put it in storage.

By this time there was only one other Maxi driver in Featherstone. On my way in to work one morning I noticed that the man with two Maxis had swopped one for a caravan. Tragically a few weeks later, his other Maxi had been shunted at the front and back. It was a write-off.

My next Maxi was sat on a quaint lawn in suburban Sheffield. I could tell as soon as I saw it that it was in his way and he desperately wanted rid of it. I offered him £65 and he took it, so I was the proud owner of my first Maxi 2. It was the worst Maxi I have had. At some point in its life it must have been in a serious accident because no matter what I tried, I just couldn't stop it going sideways. Driving it demanded intense concentration, and about three months after I bought it, exhausted following a long trip to Edinburgh, I had the engine removed and dropped into my Maroon one, which was still in storage, and scrapped the shell. It was the first time I had been forced to resort to that, but it was a nasty piece of work, so I felt no remorse.

Then my dream came true. The only other Maxi driver in Featherstone tracked me down in a local fish and chip shop. He told me he'd bought a new car and wanted to get rid of the Maxi. Would I be interested?

'Might be.' I was playing cool.

'You can have it for fifty quid.'

For thirty-five pounds I had cleaned up the town. It was my fifth Maxi. I have had two since, one was an immaculate Maxi 2, it looked and ran perfectly, but only a day after I bought it, a mechanic pointed out that the underneath bit was split down the middle. I haggled for a refund and handed it back. The last one I bought I still have. It cost me £200, the most I have paid for one, but worth every penny. It was in lovely condition. It is a wreck now of course but I think it's still got a bit of life in it.

I really don't know when this infatuation is going to end, Maxis are getting thinner on the ground as each year goes by. The Austin Maxi Owners Club do a fantastic job of keeping the old cars going and Morry Cook, the spares secretary, can put his hand to almost anything. So maybe the Maxis might make another twenty-five years.

The following section is full of Maxi love stories and people who have bonded with their car. It is nice that other people have bonded, it is nice for the Maxi to be loved. I wonder if anyone has ever bonded to a Ford Cortina? I doubt it.

Two Loves

The top Maxi belongs to Phil Malby. It is the oldest Maxi in the world and with Phil's tender loving care, is still in mint condition.

The bottom Maxi is loved in a different way, it is John Leech at Roosecote Raceway. On the North-East banger circuit, the Maxi is king.

Part Of The Family
In Less Than A Week

I'm not sure why I chose an Austin Maxi. It was back in 1982, and I was fed up with the diabolical fuel consumption and poor rearward visibility of the Range Rover, so I found myself in the market for something cheaper and more practical. We certainly went to the car auctions with an open mind, but I think a Maxi must have been on the short-list. It wasn't so much that it fitted the bill particularly well, but Maxis were universally derided - cheap, in other words - and they had a rugged reputation.

There wasn't much that caught my eye that day, but I did spot a fairly tidy blue 'J' registered Maxi booked for the very end of the sale, and with a derisory upper limit of £25 in my head I entered the bidding for a car that appeared to be running on three cylinders. Half an hour later, and £45 worse off, I found that it was not only running on three cylinders, but missing a gear as well. However, a provision of five seemed unusually generous at the time, so we drove it home firing on three cylinders and minus fourth gear.

Two weeks later, after welding and filing the cracked cylinder block, and rebuilding and reaming the third and fourth gear selector fork, I dropped the engine back in for a trial. Like magic, there were four functioning cylinders and five gears. It lasted less than a week - the engine never missed another beat, but the gearbox mysteriously lost third, while fourth suddenly needed a knack. Clearly it was knackered, as several people helpfully pointed out. But I was becoming quite fond of the Maxi. It had a special charm that was obvious only to its closest friends - outwardly civilised and respectable, but beneath the surface a willing and surprisingly gutsy tow vehicle. So the Maxi settled into a tough career in its twilight years, towing upwards of two tons without serious complaint: first gear, second gear, and an optional fourth for going downhill. They were all the gears you needed really, towing two tons behind a 1750cc engine. And it managed at least 40mpg unladen and around 25mpg towing, which was a revelation after the Range Rover. In fact, it was quite simply a better towing vehicle, unless you were in a hurry, of course. Months passed, and the Maxi plodded on with four cylinders and three and a half gears, never complaining, and without experiencing as much as a flat tyre. Occasionally things fell off, and if they were non-essential they were left off, but an ignition switch came into the essential category, so I fitted a button from a Mark 2 Jaguar.

One summer we elected to take part in the Weymouth carnival procession, and as my brother had just completed a round Britain charity ride on his Triumph motorcycle, and Triumph were knocking on the receiver's door, we chose a Buy British theme. The Maxi - I know not why - became the 'Magic Maxi', and was soon cheerfully emblazoned thus across the bonnet, with a matching fluttering Union Flag and the ominous legend, 'Buy British'. British Leyland seemed to be in terminal difficulties at the time, so we decided to do our bit and fly the flag. Naturally the Magic Maxi, which had never missed a beat in years, threw a small tantrum after three hours at 1mph, and ejected a Vesuvius of steam on the home straight, in front of thousands of spectators. Actually it was my fault because I'd overfilled the radiator to be on the safe side, and it had overflowed onto the hot exhaust, but the damage was done. The Leyland empire continued its inexorable decline, and the poor old Triumph cooperative never reopened.

The Austin Maxi was, of course, best known for its ability to metamorphose into a double bed, and I remember seeing the early 70s advertisement featuring a couple bedding down for the night with alarm clock and pyjamas, although at the time I couldn't see the point. Years later the point became very obvious when we took a Maxi-full of junk to the Beaulieu Autojumble - a three day event if my memory serves me right. On the first evening, as the rain set in, and other stall-holders began wrestling with their tents, the kindly folk to our left offered us a small tent, astonished that we intended to sleep in the car. But we had to refuse, because ten minutes later, when the other stall-holders were still wrestling with soggy canvas - we were sitting up in bed watching the television, and listening to the homely bubble of the electric car kettle. In retrospect it was probably the Maxi's finest, if rather static, hour.

And so the years passed. We threw equipment in and out of the Maxi, uprooted trees with it, turned cars over, towed trailers, taught friends to drive, slept in it, ate in it, and generally got excellent value for money from our £45 investment. Friends and relatives who had scorned the car, quietly bought Maxis of their own - just ordinary vehicles of course, quite unlike the elderly Magic Maxi, which had achieved minor legend status. And it was not destined to die quietly. I had been studying the world of motor trials for some time, wondering whether a complete novice might expect to enter, just once, and walk away with a prize. The weak spot appeared to be in the front-wheel drive classes where we guessed that a

Magic Maxi - with firm hydroelastic suspension, extra tort springs and a sump guard - might stand a chance. Being quite unsuited to hill-climbs, the front wheel drive class was generally ignored by the experts, and as the chosen event approached, we found that the opposition consisted of just three competitors: a Mini, a Citroen 2CV, and an Alfa Romeo. It looked easy, and it very nearly was.

The regular competitors in the annual Woolbridge Motor Trial gave us a wide berth on the day, fearing, I suppose, that the Magic Maxi had only been entered to make fun of the whole thing, which was more or less the case. The first few hills were easy, but we soon noticed that everyone else was adjusting their tyres to suit the conditions, and we did not have a pump or a gauge or anything else. We began to fall behind. Then the 2CV and the Alfa Romeo dropped out, leaving just us and the Mini. It wasn't to be, though. Within yards of a crucial summit the Maxi's carburettor dashpot jammed shut and we slid to a halt. The marshals were naturally astonished to see us there, within inches of the top, but it was a bitter moment. Like the stumbling industry for which the Magic Maxi so proudly flew the flag, it seemed we were destined to fail. We failed all the remaining hills as well, and were pipped for the cut-glass tankard by the Mini, so British Leyland didn't do so badly after all.

The disappointment of the hill climb seemed to affect the Maxi, and it never quite recovered its poise and - dare I suggest it - sparkle. But it was still something of a surprise after the usual pre-MOT welding and brake pipe-fettling to hear our tester announce solemnly that he would pass it once more, if we promised never to attempt another MOT. So the terminal clock was ticking away, but the Maxi struggled gamely on, burning oil by this time and flexing disconcertedly on the bumps. As the end approached, we moved house, and the Magic Maxi came into its own towing load after load the twenty miles or so to the new property, then racing back in fifth when the wind was set fair, or fourth on less helpful days. But with the final load moved, it was time for the old car to go out to graze, so we left it in the middle of a field at my parents' smallholding. There it continued to undertake light agricultural duties for a few months. Finally, as the body and subframes began to separate, I put a battery in it for the last time, pressed the Jaguar starter button for the customary half second, and drove it to the scrap yard. Bowling along in the elusive fourth gear, most of the exhaust fell off, and in a final symbolic act of defiance, the quiet middle-class car emitted a mighty bellow. I only saw it once more.

Six months later I happened to pass by, and sat one final time in the tattered seats, pressing the start button, and toying with the quixotic gear shift. Actually it wasn't a wasted visit because I came away with a long-lost spanner, a screwdriver, some underwear and an odd sock. You see, it wasn't so much a car, as a home-from-home. The best and the worst home I have ever owned.

My earliest recollection of Maxis is one of long Summer holidays in France and Spain with an uncle, an aunt and my cousins, four of us, aged 8 to 12 wedged in the back seat. With a flapping roof-rack overhead and an overstuffed bow-legged trailer groaning behind we hurtled across the continent, a Socialist lecturer and family in the British people's car, it really was the 1970s. Seduced perhaps by these memories, the first car that I bought had to be a Maxi. After all, if you could transport 6 people, a tent the size of Saint Paul's, half of Sainsbury's and enough clothes to fill the Albert Hall, to Barcelona and back, a second-hand one ought to be all right for the odd shopping trip.

I bought it for £400 in 1982, 'M' reg it was. I remember buying it now, a hideous purple colour, the owner ripped me off for another fiver by hiding one in his pocket. Well, I didn't realise that then, I didn't know that you should always count the cash out in front of the seller and best with a witness as well. Lay it on the table, get them to count it too, all agree and shake hands on it. I just gave him the money, he walked round the car and came back with a fiver short and claimed I'd tried to rob him, well, I was too green to argue, so £405 it cost really.

When we'd had it on the test drive I asked what the faint knocking was. 'Nothing,' he'd said, 'it's one of the tyres. They don't cost much.' I bought it anyway. Four weeks later and I'm doing my first engine-ectomy with mate Mark. In the end we hoisted it out on the back gate lintel. 'Big end's knackered,' I couldn't have told one from a bleed-nipple at the time so I just got another engine from the scrappers - £35 'guaranteed'. Of course, they'd sell you anything, but I knew that already. I was lucky, it worked.

Jen and I went everywhere in that car, it saw England, Wales and Scotland from end to end. Those ridiculous front seats, you really could sleep in them when you got them wound flat. The dogs slept in the back or underneath, it was better than a caravan. We only had it three years, rust killed it of course. One day I wondered why the back seat was wet, we pulled it out and looked down straight through to the Tarmac. It was too late to do anything. Four months later I got £40 back from the same scrapper that the engine came from. I didn't mind, she'd done us proud, there will never be another car like her for me.

You never forget your first love...

My Maxi story goes along the lines of: I need a cheap old car to run around London in, I buy old banger, fall hopelessly in love with it, spend very large amount of money restoring it to original condition - however, I am getting ahead of myself.

In 1990 my lady and I needed a car. We decided that we would buy an old banger given that we live in an area of high car crime and anyway did not wish to spend very much money. One evening a friend rang to say that her elderly next-door neighbour was selling his car and it seemed like the kind of thing we might want, so we went to have a look. I must admit that on first sight I was unmoved that it was a Maxi. My lady test-drove it round the block. 'Steers like a truck,' she said, but it went, and I was much taken by the old couple selling it, who had had it from new and were extolling its virtues and showing me a full service history. We did the deal for £350. As they waved us off the old lady said, 'She's called Trudy.'

I was busy at work for the next few days and it was not for a while before I could give Trudy a thorough once-over. Eventually a close inspection, and the way the car drove in general, revealed much needed doing. Then one day I opened the glove compartment and inside I found two one-dozen-boxes of hypodermic syringes with needles, and six boxes of ampules of pethadone and other heroin substitutes. Somewhat amazed, and realising that this would have been very difficult to explain to any police that had for one reason or another stopped me and found it, I put them all in a Sainsbury's bag and drove back round to Trudy's previous owners. 'I am sorry, I completely forgot about it,' said the old man, 'I used to be a heart surgeon and carried this around in case I came across a traffic accident.'

After this somewhat enigmatic start to our relationship, the more I drove Trudy the more I liked her. I liked the way the idiosyncratic changes in the way the car behaved seemed to happen on a daily basis, how it seemed to react to weather or slightly different road conditions, how journeys were unpredictable; I liked the way it drove, its lurchings and unmasterable gear lever; I liked its Tardis-like interior, the feeling I had of sitting in an armchair driving a small living room along. And I really liked the attitude you were forced to adopt when driving a Maxi, because everybody sees a Maxi coming and pulls out in front of it. You have to develop a driving style that is aggresso-defensive, and the style rubs off. I

Wives and Maxis

The top photo shows John Arnold's beautiful Maxi in a loving pose with his late wife.

The not dissimilar bottom picture is of Ian Dees' Maxi and wife. Ian wrote that he would 'rather swap his wife than his Maxi.' Admittedly it is a rather fine Maxi, but having met his wife, I'm sure Ian was joking.

became a Maxi driver with attitude, and that is a contradiction which your average Ford boy racer just does not understand, and he sure can't handle being overtaken by this Maxi - going fast.

The more I liked this car obviously the more I was prepared to spend on putting it right. I will spare you all the technical details but four years later we are up to just over £2500. I have got a Maxi in very good condition, but if I had spent this sum, plus the original purchase price, four years ago at the height of the recession, I probably could have got an excellent BMW. However, I have got to say I would far prefer to own Trudy. This attitude has to be questioned.

I have owned a number of motor vehicles, including off road vehicles, and having lived on the canals for a number of years, also a number of boats. These things are not new to me, so why get so passionate about one particular car? Firstly, I just like old cars, and Trudy is a solid mass of mild steel and chrome lacking nearly all refinements found in a modern car and from the days of 'they don't make 'em like that anymore'. Secondly I like the people you meet when you drive a Maxi. They are generally male, in their mid fifties and you have conversations with them that always start with, 'I used to have one of those,' and which then lead on to numerous anecdotes about faults, breakdowns, carrying capacity.

In 1992 my lady and I decided to have the only motoring holiday we have ever had and took Trudy on a 16 day, 2600 mile trip around the coast of England and Wales, nearly all on B roads. The car performed magnificently with no trouble whatsoever and seemed ideally suited to this leisurely pace through the country lanes. In fact, if I did not know better I would say it enjoyed the trip. We had numerous encounters because of the car, of which one example was this: We were in a car park somewhere outside of Bridlington and I could not find the way out. On my second lap an elderly gentleman gave directions. His opening line was, of course, 'I used to have one of these.'

The Maxi, of course, is not without its detractors. A couple of weeks ago I went into a specialist model shop to enquire whether any company had ever produced a model Maxi. When I asked the female shop assistant she said no, and started to laugh. As I left the shop she was bent over the counter, laughing uncontrollably. The car has had a history of bad press about its reliability, and its gearchange is the stuff of motoring legends, although apart from the gearchange, which I have to agree with, I have found my Maxi to be no more unreliable than any other vehicle that I have owned. It is also vilified for its looks. Now I happen to teach design at a local university and as you can imagine I have at times had to put up a robust defence of my choice to drive a Maxi to my colleagues and students who feel that I, as an 'expert', should perhaps know better. I do know that it has the aerodynamics of a brick, but that is judging it by contemporary criteria, and it was a very innovative design for its time. Bearing in mind that we are talking about a 25 year old design concept it actually fits very well in that time frame. It was not a compromise of design, it was conceived to be a roomy family car, almost utilitarian in its usefulness, with plenty of ease of access for the whole family, and it fulfilled that objective. It has to be seen in the context of late sixties industrial design of its type, and in this way the very good reasons why the concept came into being can be understood. I, personally, happen to like its shape, both profile and head on. It sits very well, has a terrific product integrity and is unmistakably a motor car. Or perhaps I am biased.

The Maxi was reasonably reliable, only letting us down on a few occasions, and then doing so in rather convenient places. Once the steering wheel caught fire. Fortunately we were only about two hundred yards from home, and the fire went out when I switched off the ignition. The fault proved to be in the direction-indicator switch and the garage said that it had never come across such a fault before. It seems odd that the half a dozen cars or so that we have owned have had faults never seen before.

The water pump failed whilst we were on holiday in a remote Cumbrian village. Fortunately the place had a garage whose owner believed in service and he sent a car to Carlisle, some thirty miles away, to fetch a replacement and then fitted it, all for a very modest sum.

The next failure was also in a remote village, this time in North Wales. We drove in to its garage to refuel, and when I tried to start the engine there was no response at all. The battery had suffered instant and total failure. It couldn't have happened in a better place - the garage had a stock of pre-charged batteries and we were on our way again in no time.

The last failure was odd. We stopped for traffic, and when I wanted to move off again the car did not - nothing happened at all. Again, it could hardly have happened in a better place, we were out of the main stream of traffic and right alongside a phone box. I called the RAC, who arranged for the car to be taken to a local garage for repair. I had expected to be without a car for several days and to receive a very large bill. In fact, we had it back within the hour, together with a small bill, and the news that the driveshaft had become uncoupled due to a fall in the pressure in the suspension system. All they had to do was to put it back again and pump up the suspension.

The car did have some strange features, the hydro-elastic suspension for one. Although comfortable, the car didn't half dip when the brakes were applied sharply. I shall never forget the surprised expression on the face of the parking attendant on an Irish ferry when he slapped the bonnet to tell me to stop and found that it retreated by a couple of inches. I think he thought he'd broken something. It was on this ferry that I greatly surprised other passengers by entering the car via the tailgate because I could not open the side door far enough. I don't think that some of them had seen a five-door car before.

The brakes squealed distressingly and there was a strange squeak from the front when we turned sharply to the right. Neither effect could be cured and we just had to ignore the stares from passers-by. All in all it served us very well. We felt very sad when we saw it being driven away after we had sold it, it was like losing an old friend.

In the very beginning there was a Minivan. It came with my wife. After a while she persuaded me to learn to drive it rather than be chauffeured in it, and so I became used to the front wheel drive, transverse engine style of motoring, and liked it. Our next car was a Mini Clubman estate, but fearful that one day one of our elderly relatives would become permanently lodged within the back of such a confined space we progressed logically and naturally to our first Maxi, and the elderly relatives loved it for its comfort and ease of access.

The novelty of having five gears brought a dramatic moment to our very first test drive in a Maxi. Having at last got onto the open road I slipped into the fifth gear, no problem, and happily cruised along for a while, but eventually turned off down a slip road leading to a roundabout. At this point, whilst still moving fairly briskly, I changed down - straight into second gear - proving yet again that I'm better on theory than practice when it comes to driving techniques. The car coped better than the accompanying salesman!

In those days versatile cars were still a rarity. If you wanted lots of storage space you had a full blown estate or at least a van, so I always felt a little smug on my frequent visits to the local timber yard when, amongst all the other weekend carpenters trying to persuade their nine foot lengths of three-by-two to fit in through the sun roof or out of the side window, I just opened the hatch, slid in all the timber I needed across the single bed, shut the hatch and drove away without a single dangerous projection or draughty opening.

That first Maxi served us well, but after seven years and seventy five and a half thousand miles of trouble-free motoring, we decided to trade up to a twin carb HLS. This was not a happy experience, and more correspondence was generated between myself and British Leyland and the garage than were miles driven.

The greatest ignominy it brought upon us was an incident in France, when a horrendous noise from the front forced us to abandon the car by the roadside, and to walk into the next village to seek help via the gendarmerie. They obviously thought we had just run out of petrol, but my description of 'un bruit horrible' convinced them that serious mechanical help was needed. 'Quelle est la marque de votre voiture', 'Austin Maxi '. You could tell by the look of the officer's face this was not going to be easy. However, they made some telephone calls, and several hours later it found itself up on a garage ramp, whereupon the garage proprietor discovered a loose nut that had fallen off and wedged itself next to some moving part - hence the noise. The nut was removed and refitted where it belonged and 'Voila - elle marche'.

Refusing to keep that particular car beyond its warranty period we exchanged it for a straight forward HL. But during the year the Maxi had had its final face lift and had become the Maxi 2. So more by accident than design we ended up with this final very attractive version, and now some fourteen years on it's still as smart as when it first came out of the showroom. Now it attracts many admiring glances, when originally, as just another boring new Maxi, it was disdained by the casual observer.

All our three Maxis have been owned by my wife, even when they were our only car, but now UPU 4OW is very much my wife's personal car and, as a consequence, it leads a leisurely, well cossetted existence, but in its prime it certainly earnt its keep. It regularly took us to Switzerland and Scotland, in winter as well as summer. Its spacious rear helped to shift beds across the borough, it served as a sentry box for ticket sellers at the school gate for Bonfire night, and most patriotically, in its bright red livery, it acted as chuck wagon at the Street Party to celebrate the Royal Wedding. Ah, that all British institutions could last as well as the Maxi.

When Austin Rover or British Leyland, or whatever name they were then operating under, stopped making the Maxi, the local dealer who had supplied my father with a new Maxi every two years for ten years had the audacity to suggest that... wait for this... the Austin Allegros would be a worthy replacement vehicle! Suffice to say, my father was not convinced, and truth to tell, a Golden Era had passed. The Maxi had served our family well whilst we children were growing up, but now we were beginning to make our own way round the countryside, we didn't really need the early 'people-carrier' that the Maxi was anymore, except for the annual family holiday with a ludicrous number of surfboards in the back. Father bought a Vauxhall Cavalier after that and things were never really the same again. The two of us who had to sit on the outside position of the back seat always had to bend our necks to one side, because of the curve of the Cavaliers roof; something to do with aerodynamics...it would never have happened in the Maxi!

I think that this car must be runner up in the world championships for attracting ill-informed, stupid and prejudiced journalism. I have an article taken from the magazine section of one of last Sunday's papers. This journalist must be one of the leading exponents of the idea that truth should not be allowed to stand in the way of crowd pleasing journalism. His article is simply ridiculous. Incidentally I think that the holder of the world title must be the Allegro - of which I have also had several.

My wife and I have both been driving for over 40 years with more than 25 cars and we have had mainly, but by no means exclusively, BMC/Leyland/Rover cars. Whatever else one might say about those cars, they have always been extremely reliable. During the 70s and 80s it was fashionable for sections of the motoring press to criticise British cars for reliability; *Motoring Which* was one particular culprit. I never could understand the basis of such criticism since it was so contrary to my own experience, but I had to accept that my experience must have been different to the majority. No doubt that unfavourable press did much to bring about the collapse of the British motor industry.

Nowadays I read in the classic car magazines about the legendary reliability and durability of these old Austin engines and gear boxes! I sometimes wonder why the national press seemed to have a death wish for the UK industry, or was it just fashionable, the 'in thing', to knock them? I am sorry that I cannot tell you about the time my Maxi pulled a broken down truck halfway across the Sahara desert, or how I managed to get twenty people in it whilst we were fleeing from this or that trouble spot. The truth is that both cars fulfilled a normal mundane family car role of commuting and leisure activities, but they did perform it in a near faultness manner.

In 1973 I joined Pye Unicam Limited, the Cambridge based laboratory instrumentation wing of the Pye/Philips electronics empire. I worked as a senior field service engineer on a range of UV/visible, infra red and atomic absorption spectrophotometers in the London, Kent, Surrey, Sussex and Middlesex area. In those halcyon days I covered an average of 35,000 to 40,000 miles per annum and received a new Austin Maxi each year. I left the company in 1980.

The Maxi was a very underrated car, for which I hold many fond and at times, bitter memories. It was truly an all purpose vehicle - a work horse, a comfortable tourer, tow car and a double bed. My first Maxi was a 1500cc in a disgusting mustardy khaki colour. Within nine months I had progressed to a coveted 1750 in Teal blue. What a difference! Good-looking, responsive and a pleasure to drive.

There was the time I tried to obtain two replacement grommets for the unused windscreen wiper holes to be told by the receptionist that I had recently visited a safari park. I was amazed by his powers of perception. I had been to Longleat House and Safari Park the previous weekend. He went on to tell me that the grommets had most probably been removed by the monkeys that climb on every car passing through the park. Maxi grommets were obviously a treasured possession!

Another experience was the failure, through metal fatigue, of the accelerator pedal return spring. There I was, lazily driving along at about 50-60 mph, when bang, the spring snapped and the car got catapulted into the future as the engine raced to its inevitable destruction.

Of course the only way of preventing this catastrophe was by turning off the ignition. As the car slowed down, a feeling of calm would start to return until the grim realisation dawned that I had turned the key back too far and engaged the steering lock. Ninety in the fast lane of the M2 without the ability to steer; not a healthy pastime!

I used the car on many holidays, with and without a caravan, and during the seventies toured Scotland, Wales, the Lake District and the South West without a hitch, except on the following occasion.

During a holiday in the South West, the rear wheel bearings started to make the most appalling noise and so I took the car to the nearest dealer. Apparently the original bearings had never been packed with grease and replacements were immediately fitted. Unfortunately, in their rush to relieve me of some money, the dealer omitted to mention that he had fitted Mini bearings in place of the originals. Same diameter but only half the width; they lasted for about 500 miles.

On another occasion, my Maxi had just been serviced. I drove away happy in the knowledge that my local dealer only employed professionals. Not a bit of it! They managed to 'install' an electrical fault which proceeded to heat up the choke cable. This in turn set light to the fibrous sound-deadening material behind the dash board. I stopped, removed handfuls of burning material and managed to save the car from its premature fate.

Unfortunately, I received some minor burns which took a month to six weeks to heal! Never again. Next time I'll enjoy the pyrotechnics!

My last Maxi, before leaving the company, was a black 1750 with a beige interior. It looked like a million dollars! I even customised it with a smaller steering wheel - a useful alternative to the standard wheel obviously purchased from their coach depot in Leyland - and a set of wheel nuts and hub trims to replace the, by then, old fashioned looking mirrored caps which were still being fitted. No furry dice I hasten to add!

To summarise, I think the Maxi was probably one of the most innovative cars of its time. I am still a company car driver and have covered 35,000 to 40,000 miles a year, every year to date, in a variety of prestigious makes. However, I doubt if any of them have given me the pleasure -perhaps it has something to do with the fact that I was in my twenties then - or pain, has the Austin Maxi 1750 did.

I've been driving Maxis for twelve years, I have had four in all. For me it is the best value for money car that there has ever been and I can't see myself driving any other car. I'm getting on a bit now and I know I won't live forever. At my age you start thinking about your funeral and how you would like it to be. Well, I bet you can guess what I'm going to say next. I want my last journey to be in a Maxi, I'm sure that with the seats folded back, there will be plenty of room for my coffin.

My grandson drives about in my old car at the moment, I taught him to drive in it and he used it on his driving test. He failed the first time because he had a puncture, but we got it fixed and he passed second time around, no problem. I can see that he is becoming as fond of the old Maxi as I am, so I let him drive it whenever he wants. It's a lovely feeling. There is something grand about your offspring following in your footsteps, or in this case, following in my skid marks.

Long live the Austin Maxi.

Epilogue

I have had a lot of fun working on this book. To be fascinated by something for so long in isolation, then suddenly to receive piles of information and stories throughout the course of one year is very exciting.

There are still vast amounts of Maxi stories out there that I haven't got in full enough detail to include. I wish I had more information about the North East banger circuit, where the Maxi is very prominent and is the car to beat. I wish I could have contacted people involved in the 1970 World Rally that got so much attention and included Prince Michael of Kent in a Maxi. Then there are the stories that I only got snippets of; a women rang me and claimed she was a direct descendant of a Russian Prime Minister from the late Nineteenth Century and she travelled all over in the Maxi, including an epic chase through the desert where she was pursued by gun toting Arabs, but I never heard any more from her. Who knows there might be another book out there.

So that's it, the end. As to *Why? Twenty Five Years Of The Austin Maxi*, the answer is I reckon the attraction of the Maxi is that it can be whatever you want it to be. So many diverse people seem so attracted to the car that I presume that the Maxi must fit into their own personality rather than have a distinct one of it's own. It can change it's colours at will, it is a chameleon on wheels.

I will go now, climb into my Maxi, get to the end of the road and breakdown. Of that there is no doubt.